JUMPING FROM THE IVORY TOWER

Weaving Environmental Leadership and Sustainable Communities

Rosemarie Russo

University Press of America,® Inc.
Lanham · Boulder · New York · Toronto · Plymouth, UK

4501 Forbes Boulevard
Suite 200
Lanham, Maryland 20706
UPA Acquisitions Department (301) 459-3366

Estover Road
Plymouth PL6 7PY
United Kingdom

Printed in the United States of America
British Library Cataloging in Publication Information Available

Library of Congress Control Number: 2009938786
ISBN: 978-0-7618-4980-3 (paperback : alk. paper)
eISBN: 978-0-7618-4981-0

∞™ The paper used in this publication meets the minimum requirements of American National Standard for Information Sciences—Permanence of Paper for Printed Library Materials, ANSI Z39.48-1992

This work is for my son, Kyle and all future generations in hopes that the wilderness will shape their potential to walk softly on this planet. It is for Steven Hunter, who came to me during the first week of this journey. I think of you on every first run of a steep slope. I trust you have found peace in nature.

CONTENTS

Preface

Dr Rosemarie Russo
Fort Collins, Colorado
June, 2009

Traditional college-level environmental curricula do not always provide the basic skills needed to address the environmental issues today such as mass extinction, climate change, loss of biodiversity, habitat destruction, and others. Despite the increasing number of studies on service-learning, environmentally oriented college-level service-learning research is limited. My experience after twenty years in the education field is that service-learning holds the key to changing behavior not just minds. As a nation we are not producing environmental leaders that can guide us toward a sustainable society. This book examines the integrative potential of service-learning in regards to skills and behavior deemed critical by environmental educators. This research explored the correlations between scientific skills and affective behavior, an area unexplored in the majority of college curriculum. The work also identified environmental education deficiencies and methods to address those limitations. Narrative analysis ascertained gender preferences related to a collaborative approach in the sciences, which illustrates a simple method that could be used to increase the number of women in the sciences.

Acknowledgements

To Courtney, Jared and Annette, your dedication and clerical support allowed this poor inner-city kid to reach the highest education challenge. A special thanks goes to my editor whose work as a legal scholar, nature writer and researcher has not only inspired the work, but has assisted in translating the research into academic prose. To Kook, your 90th birthday present is having a granddaughter who is now a doctor. To my family and the Bean family for their babysitting and encouragement. To Carol, Anne, Tom and Jennifer, who set the bar and entertained me at the bar while I conducted my research. To Dr. Jessica Young, George Silbey and the Rutgers and Western State students, thank you for participating and for protecting the Gunnison and Raritan Valley. To the cover artist that created this woven piece that symbolism the connections that are needed to preserve the planet.

Cover artist: Jennifer J. Wilhoit is a published writer and story-gatherer with a passion for how we are made whole through our connection with Nature and the Creative. In addition to personal essays, she listens to and documents others' soulful stories about spirit, creativity and nature through her endeavor, *TEALarbor stories*. When she isn't writing, story-listening, traveling or spending time in the natural world, Jennifer does her own creative work such as the handwoven watercolor collage shown here.

Introduction

In 1985, I began a teaching career as an Oceanographer and Environmental Educator, throughout the course of my career I realized empowerment techniques are missing from the majority of textbooks and nearly all classroom assignments. Therefore, my pilot studies and dissertation research, which combined grounded theory and action research, explored innovative research techniques for improving environmental decisions in terms of inclusion and empowerment. I took lessons from that work to write this book. The book is meant as a resource about why educational instruction needs to be modified and shows the positive results of linking real world projects with academic lessons to produce individuals that not only can change the world but "want" to make a difference because they developed a sense of place.

Chapter One covers research about, the social and moral imperatives as educators to create sustainable communities and help students connect to their communities and the literature to date about the effectiveness of service learning. Chapter Two provides an overview of service learning pedagogy and a discussion of various educational models. The research examined if the traditional educational model of information-dissemination hinders students' ability to both believe that they can contribute to the community at large and participate willingly in community projects. In the traditional lecture-driven, assessment centered model, students do not learn how to apply knowledge to real world issues. The epistemology of service learning and its foundation base within the experiential and constructivist paradigms is examined. The review analyzes the positive and negative service learning claims, service learning within the broader context of education and specific student outcomes.

Chapter Three describes the politics of science including traditional ecological knowledge, environmental justice, eco-exclusions and college and community interactions. The chapter consists of an analysis of student's reactions to three service learning projects at Western State College and Rutgers University specifically how the experience affects collaborative learning, enhancement of subject matter, exclusive emphasis on science and environmental ethics.

Chapter Four outlines the case study findings. Chapter Five discusses how environmental education curriculum could be enhanced and examines gender preferences. The summary, conclusion, and recommendations are detailed in Chapter Six.

Chapter One
Environmental Consciousness

In the end we will conserve only what we love, love only what we understand-and we will understand only what we are taught.

Baba Datum

Environmental education is multifaceted and involves environmental economic and social consequences of human activities.[1] Environmental education should address the political, social and economic relationships that generate and sustain the problems.[2] Based on results of industry leaders and government officials' responses, environmental graduates are deficient in four main areas: civic responsibility, technical skills such as sampling and monitoring, complex thinking (i.e., dialectical & holistic) and an understanding of the regulatory dimensions of scientific work.[3] In addition, environmental educators such as Orr stress that environment science curricula should join students, teachers and community members in a pattern of connectedness, responsibility and mutual need.[4] Can service learning enhance environmental education? The critical difference and distinguishing characteristic of service learning is its reciprocal and balanced emphasis on both student learning and addressing community issues that are otherwise unmet.[5] For example, environmental students can provide their technical expertise (e.g.., water sampling techniques) on community projects and incorporate local ecological knowledge back into their academic programs.[6]

Figure 1 illustrates the complex environmental attributes examined in this study.

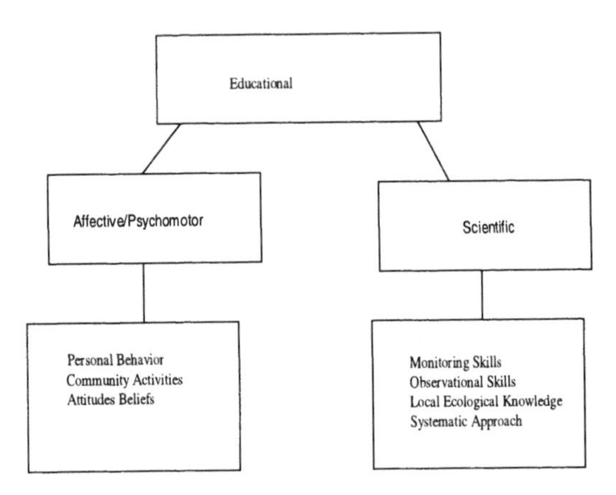

Figure 1: Education Impacts of Environmental Service Learning

Service learning is especially appropriate in the context of environmental education. The environmental challenges that all communities face provide unique opportunities for students to apply their skills in the form of community service. Moreover, service learning can enhance a deeper understanding of place. As Weil has said, connection to place is "the most important and least recognized need of the human soul."[7] By incorporating service learning into environmental curricula, students can bring community members together to develop trust and build a local capacity for renewal and growth[8].

As earlier studies show by having individuals make a commitment to a cause it creates a high level of environmental concern and subsequent constructive action.[9] The emerging theory can inform administrators and faculty about the integrative potential of service learning.

According to Miller, early studies on student outcomes revealed that students, who participated in service-learning projects: showed improvement in participant knowledge in the areas most directly related to the field experience and in the integration of theory and practice.[10]

According to Kellogg, service-learning assignments with their concrete, real world component, help beginning students see a discipline in a new light.[11] In some disciplines, the traditional theory-oriented introductions do not excite certain groups of students such as women and students of color.[12] Latina and African American women do value community work and view it as an extension of their role as mothers.[13] Service learning could stimulate their interest in environmental science by combining their community work with academic sciences courses.

Research on sex differences indicate that girls and women have more difficulty than boys and men in asserting their authority or considering themselves as authorities[14]; in expressing themselves in public and in fully utilizing their capabilities.[15] This research showed that female students felt more empowered through their experience.

Service learning can take students out of their academic shelter and help them begin to develop a sense-of place-something that is particularly important for developing the stewardship needed to protect environmental quality.[16] The research explored whether students express more knowledge and interest in local stewardship after their service-learning experience.

The knowledge about a particular watershed or community clean-up site could empower students and community members and, therefore, help them to realize that they can make a difference in environmental protection.[17] The recognition of local ecological knowledge may broaden the participant's cultural views, which could indirectly aid them in resolving disputes that have an environmental justice element. Environmental education tends to focus on either science or policy. Yet, Doughty notes, "By working in their communities students will increase their understanding of community and ecosystems.[18] They will restore their watersheds and rebuild a sense of place and belonging." Overall, service learning has the potential to enhance social interaction that may result in change and empowerment in the community.

SIGNIFICANCE TO EDUCATION FIELD

The study focused on students' perceptions of how service-learning compared to traditional classes in terms of their scientific skills, learning styles (i.e. collaborative vs. competitive), desire and ability to foster community environmental change after participating in existing and new community-based environmental educational projects. Figure 2 illustrates the main categories of areas examined through an analysis of students' narratives.

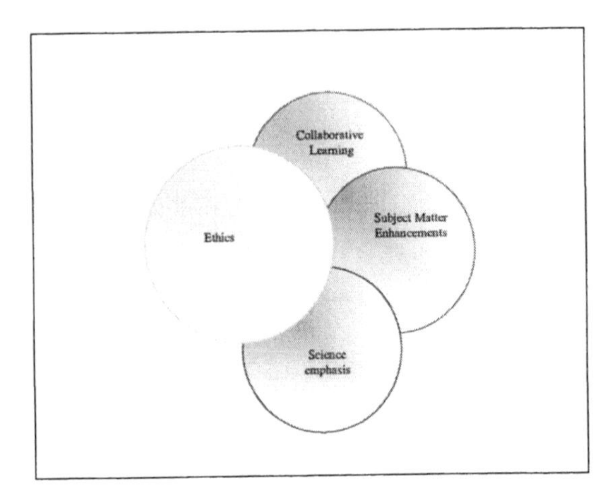

Figure 2: Service Learning Impacts

The research addressed the following questions:

1. Can students benefit by working collaboratively on local environmental issues?
2. How can service-learning enhance environmental education specifically the exclusive emphasis that environmental education places on science?
3. Does service-learning enhance subject matter learning?
4. Does service-learning contribute to the development of an environmental ethic of caring?
5. Are there gender differences in the impacts that students express from the same experience?

In regards to the question about the development of an ethic of caring, the students were questioned before and after their experience using a Likert-scale to determine if their involvement changed their attitude towards environmental protection/planning within their community.

Social and Moral Imperative

The epistemology that under grids traditional pedagogy is positivistic, in conflict with communal ways of learning, and most importantly does not advance the student's commitment to the greater good.[19] Fundamental issues related to the role of higher education in fostering socially responsible and caring citizens are concerns raised in recent social and educational analysis.[20] Boyer, the past president of the Carnegie Foundation for the Advancement of Teaching, said, "A basic feature of the social and moral imperative of education is to help students see the connectedness of all things – social, personal, religious – to the past, to the natural, and to the eternal." [21]

To answer the question, "What type of education is needed for today's society", one must examine the fundamental question, "What is the role of education?" Dewey and Freire, two of the most influential educational theorists, seminal philosophers and practitioners, spent their careers addressing that particular question. Dewey's basic premise is that the fundamental purpose of knowledge is to improve human welfare.[22] Dewey began to question the aristocratic and antidemocratic traditions of education that had existed since Plato's era.[23] Uncompromisingly aristocratic and antidemocratic, Platonic thought has had perhaps the greatest and most pernicious impact on Western education.[24] In contrast to the Platonic view, Dewey explicitly promoted a democratic educational philosophy, which is key to the foundation of service learning. Dewey argued that traditional education created "passivity of attitude," with devastating effects on a student's ability to become an active participatory citizen.[25] Dewey emphasized experience as a tool for learning. Many service learning practioners believe that education should be related to democratic values and civic participation. They view service learning as a method to engage students in public dialogue and get them concerned about their own issues.

Another educator Freire, questioned the educational status quo after he began helping South America's poor acquire literacy skills. Freire redefined the political meaning of education.[26] For Freire, education has the potential to be liberating, and liberating education is the path to knowledge and critical thinking. His life work revolved around using education as a tool to evaluate and transform the dominant social order. His ideological orientation is expressed through his demands for educational institutions to promote "a willingness to enter into dialogue with the dispossessed in society; unpack dominant myths embedded in our socialization; and examine power and class relations."[27] The development of the skills such as "entering into a dialogue" and "unpacking dominant myths" are often neglected within our current educational programs.[28] The ability to enter into a dialogue is important in various disciplines but it is critical in the environmental field.[29] Thomashow claims, "Environmental practitioners must know how to communicate and collaborate with other people, otherwise their content knowledge will be on no use to the community."

Why should our institutes of higher education take heed to the theoretical principles espoused by Dewey and Freire? Returning to the tenets of Deweyan and Freirean philosophies, both scholars share in an educational framework that promotes the theory and practice of active, democratic, reflective, and experiential education. Yet their theoretical foundations differ because Dewey is grounded in a pragmatic philosophy that focuses on the connections between knowledge and democracy. Consequently, Dewey believes that "educational institutions should be equipped to give students an opportunity for acquiring and testing ideas through a problem-posing and solving curriculum."[30] Dewey places an emphasis on grounding education within a frame of reference. He insists, "The conception of education as a social process and function has no definite meaning until we define the kind of society we have in mind."[31] "The kind of society we have in mind" could be an environmentally sound society. This topic will be revisited in the discussion about environmental education. The question then becomes how colleges can work with communities to create sustainable communities.

Freire's work on viewing society in terms of power dynamics plays into the work because citizens often feel powerless in their environmental struggles, such as battling government agencies and transnational companies.[32] Pollution exposure and hazardous waste (i.e., Superfund) clean-up projects within their communities are examples of the power imbalances.

Freire's critical pedagogy work promotes action, critical consciousness, and praxis by students and underscores fostering individuals who participate in social change. Freire's work bears relevance for environmental education because often the issues of power and control consume environmental policy. For example, the emerging field of environmental justice speaks to the issue of environmental power imbalances and racism.[33] A clear understanding of what Freire refers to as "consciousness" is indispensable for environmental education and policy. Freire describes the ethic of conscientization as:

> Depth in the interpretation of problems; by the substitution of causal principles for magical explanations; by testing of one's own findings and openness to revision...; by refusing to transfer responsibility; by rejecting passive positions; by soundness of argumentation; by the practice of dialogue rather than polemics...; and by accepting what is valid in both the old and new.[34]

The social, political and scientific fields are interlinked. Writers such as Eckerley see the necessity of merging these areas. One method to incorporate these principles is the praxis of environmental service learning.[35]

NOTES

---

1. Summers, 2000
2. Kellog, 1999
3. Ohio Board of Regent Application, 1999
4. Orr, 1993
5. Cleary, 1992
6. Hudspeth, 1999
7. Weil, 1971
8. Lisma, 1998; Miller, 1997 & Rothman, 1998
9. Pardini & Katzev, 1983
10. Conrad & Hedin, 1992; Hamilton & Zeldin, 1987; Markus, Howard, & King, 1993.; McCluskey-Fawcett & Green, 1992
11. Kellog, 1999

12. Fox & Ronkowski, 1997
13. Naples, 1992
14. Clance & Imes, 1978; Cross 1968; Maccoby & Jacklin 1974; Piliavin 1976; West & Zimmerman 1983
15. Treicher & Kramarae, 1983
16. Diffenderfer, 1999
17. Mumford, 1970
18. Doughty 1996
19. Palmer, 1990
20. Barber, 1992; Bellah, 1991; Boyer, 1990; Coles, 1993; hooks, 1994; Rhoads, 1997; Tierney, 1993
21. Boyer, 1981
22. Haravy, 1998
23. Dewey, 1916
24. Harkavy, 1998
25. Harkavy, 1998
26. Freire, 2000
27. Deans, 1999
28. Deans, 1999
29. Thomashow, 1996
30. Dewey, 1980
31. Dewey, 1980
32. Tanner, 1972
33. Lavelle, 1992
34. Freire, 1973
35. Eckerly, 1992

Chapter Two
Environmental Education Enhancement

When you find your place – where you are
learning occurs.

Chapter Two focuses on several salient issues: educational theory, traditional pedagogical model of education, service learning epistemology and praxis, experiential learning, constructivist paradigm and environmental education. The educational philosophy of the prominent educators such as Dewey, Freire and Greene are discussed. Although service learning as pedagogy is not explicitly cited in their works, it certainly is at the base of their teachings.[1] The educational philosophy and praxis of these scholars place the ideals of service learning within a historical context. The second part of the chapter examines studies of service learning outcomes across different academic disciplines, mainly from the perspective of the student. The literature on specific college-level environmental service learning is rather limited.[2] Deficiencies in the literature are noted. Environmental advocacy is the third major area of inquiry. Finally, this section also compares information dissemination and service learning epistemology in terms of student outcomes.

Traditional Pedagogical Model of Information Dissemination

In the traditional information-dissemination model, learning is individualized and privatized[3]. It is focused on performance and curricula-related information. In traditional information-dissemination courses, faculty control is important so that the material can be transferred to the student in an efficient manner. The high degree of structure and control provided by the instructor leads to a passive learning posture by the students. [4] The information-dissemination model is geared toward preparing students for tests, however, classroom practices specifically designed to prepare students for tests do not foster deep learning .[5] Other studies describe the discrepancy between perceived and actual student success as the difference between learning and performance.[6] Furthermore, according to Katz, "the emphasis on performance usually results in little recall of concepts over time, while emphasis on learning generates long-term understanding".[7]

According to Greene, the focus on assessments hinders deeper learning.[8] An American philosopher of education, Greene poses the question: "How can we justify a commitment to critical reflection, aesthetic awareness, open-ended growth or intercultural understanding to a public occupied with the need to focus on the skills and proficiencies alone." [9]

In the traditional information-dissemination model, most teachers rely heavily on textbooks.[10] The information that teachers disseminate to students is directly aligned with the information offered by textbooks, providing students with only one view of complex issues, one set of truths .[11] The fragmentation of the curriculum and the pressures of time have made intellectual inquiry so highly specialized that by the 7th grade, most curriculums are departmentalized and heavily laden with information to be memorized .[12]

Education needs to move beyond the traditional model to provide a foundation for individuals interested in social change. [13] The institutions themselves need to become engines of change.[14] Contemporary educators have approached this type of reform in a number of ways: citizenship school; [15] the banking and problem-posing models of education; education as a spiritual journey; and education as the practice of freedom.[16]

EDUCATIONAL FOUNDATION AND SERVICE

The roots of service in this country go back to early U.S. history, when Alexis de Tocqueville described certain acts of civic and social support.[17] Tocqueville saw these "habits of the heart as a counterpoint to the individualism in American society, and as a way to unify the political community and support free institutions." [18] This sentiment is echoed by contemporary educators such as Hornbeck, who promotes the use of community service to improve education.[19]

Freires' critical pedagogy in relation to action, critical-consciousness and praxis underscores the notion of fostering an individual who participates in social change. The question is how we educate students and community members. Freire's work is particularly relevant for the environmental field because often the issues of power and control consume environmental policy.[20] The emerging field of environmental justice in education and the legal field speak to this issue. Former President Clinton issued an Executive Order on Environmental Justice in 1994. No state legislation has been passed with the exception of state laws in Arkansas and Louisiana to adequately address environmental justice.

Grassroots' leaders have emerged from groups of concerned citizens (many of them women) who see their families, homes and communities threatened by some type of polluting industry or government policy.[21]

Often times the issue of racial bias or "perceived" racial bias tends to create adversarial situations that do not separate the people from the problem. A growing body of evidence reveals that people of color are subjected to a disproportionately large number of health and environmental risks in their neighborhoods (e.g. childhood lead poisoning) and on their jobs (e.g., pesticide poisoning of farm workers).[22] In order to resolve environmental disputes, one needs to re-evaluate the complex cultural, scientific, economic, legal, and power issues surrounding environmental disputes.[23] Choucri explains, "One of the most important environmental achievements is the building of consensus between scientists and policymakers in the development of a flexible framework designed to avoid obsolescence in the face of new scientific evidence.[24]

SERVICE LEARNING EPISTEMOLOGY

Is our traditional pedagogical model serving our students and society? The service learning pedagogy departs from the traditional faculty-focused, lecture-driven model to a focus on a participatory approach to education. In service learning settings, "learners develop more inclusive, differentiated, permeable, and integrative meaning perspectives. In doing so they move toward higher levels of reflective judgment and higher orders of consciousness." [25] According to Harkavy, one of the pioneers of service learning, "the notion of academically based community service started to develop out of a participatory action research mode, asking the question of how universities can change the world".[26] Service learning epistemology evokes a fundamental change in both the philosophy and pedagogy of education. The roots of service learning emerged from the experiential and constructivist paradigms. Experiential education centers on placing the student in a real world environment and encouraging a hands-on approach to learning. In a constructivist classroom, the teacher searches for students' understandings of concepts, and then structures opportunities for students to refine or revise these understandings by posing contradictions, presenting new information, asking questions, encouraging research, and or engaging students in inquiries designed to challenge current concepts.[27]

Constructivism centers on a holistic approach to education based on a student's interest. Constructivist pedagogy lends itself to service learning because students are empowered to choose projects of relevance to them in conjunction with meeting a community need. It allows them to use their creativity to ameliorate and reflect on real community challenges and create new understanding of the world around them. Noddings, a moral philosopher, claims the premise of constructivism is that:

> It recognizes the power of the environment to press for adaptation, the temporality of knowledge, and the existence of multiple selves behaving in consonance with the rules of various subcultures.[28]

Both experiential and constructivist theories readily provide the foundation for service learning.[29] Service learning programs place students within different environments, among different stakeholders, which provide a setting for new constructions of knowledge. Although based upon experiential and constructivist methodologies, service learning advances the process to a more comprehensive level.[30]

Service learning is closely linked to the theories of experiential learning and constructivism, below is a brief discussion on both.

EXPERIENTIAL EDUCATION

Experiential learning, often referred to as reflective learning, is characterized by the following principles: learning is active rather than passive with lectures and recitation used to enrich understanding; experimentation plays a major role; risk-taking is encouraged; and the results are evaluated based on goals set by the learner and the instructor working together as a team.[31] Through experiential learning, students make their own choices and learn to accept the risk of failures. There is a tradition of learning through experience in public education.

The pedagogy of experiential learning endorsed by Dewey promotes a continuous and adaptive learning approach. Dewey believed that education should be a "process of living" which fosters meaning and comprehension through concrete experience, reflective observation or assimilation, abstract conceptualization or theory building, and active experimentation or problem-solving.[32] Dewey presented the importance of reflection in learning, and Kolb synthesized Dewey's point in a model of how the experiential learning process works.[33] Kolb provided a conceptual framework on how experience and learning connect. Interestingly, reflective thinking has become the operational linchpin of successful service learning pedagogy.[34] [35] [36] [37]

Kolb's model presented below outlines the critical conditions for the learning process. These steps are similar to the service learning process in which students will be required to engage in an activity, write and reflect upon their activities, link their field activities with course content and comment on any new commitment that they may experience after the project is complete. Figure 3 shows the process the students followed: they conducted three projects, observed in the field, keep field notebooks, tied fieldwork with lecture about ecological resiliency and habitat destruction and finally made recommendations to local and state officials based on their active experimentation.

Active Experimentation
Testing, new ideas,
decisions, commitments

Concrete Experience
Service activity

Abstract Conceptualization
Linking to course content,
theory, new paradigms

Reflective Observation
Journals, writing,
discussion, activities

Figure 3: The Learning Cycle adapted from Kolb

CONSTRUCTIVIST PARADIGM

The constructivist pedagogy, a method that examines how people come to know their world, centers on the learner's ability to internalize and transform new information. As articulated by Fosnot, "Learning is not discovering more, but interpreting through a different scheme or structure."[38] Bruner describes the constructivist paradigm as a technique of inquiry:

> It is the hunch that it is only through the exercise of problem solving and the effort of discovery that one learns the working heuristics of discovery; the more one has practice, the more likely one is to generalize what has been learned into a style of problem solving or inquiry that serves for any kind of task encountered – or almost any kind of task.[39]

Bruner's analysis is consistent with Piaget's foundational theory and the operating premise of constructivist teaching.

Piaget, one of the most influential proponents of constructivism education, stated, "The growth of knowledge is the result of individual constructions made by the learner. It is a process of continual construction and reorganization". In addition, "the human mind is a dynamic set of cognitive structures that help us make sense of what we perceive." [40] These structures grow in intellectual complexity as we mature and interact with the world. A constructivist teacher seeks the students' points of view in order to understand their theories about the world and use their present conceptions for subsequent lessons.[41] Teachers take on the role of mentors guiding the learning, not dictating it.

The constructivist vista involves creating an atmosphere in which students are encouraged to explore and think critically. Service learning curricula embrace constructivist principles. According to contemporary educators, constructivist teaching:

> Frees students from the dreariness of fact-driven curriculums and allows them to focus on larger ideas. It empowers students to follow their own interests, make connections, reformulate ideas, and reach unique conclusions. It provides statements with first-hand knowledge that the world is a complex place in which multiple perspectives exist and truth is often a matter of interpretation. It lets them see that learning and the process of assessing learning can be a complicated process.[42]

The majority of service learning research has focused on impacts to students' academic, career, civic, personal, ethical and social development.[43] Kahne notes, "There is a need to research the impact of service learning on the civic purpose of higher education.[44] Zlotkowski advocates for studies that help us understand how service learning fits with the norms, perspectives, and basic assumptions of the various academic disciplines.[45]

COMMUNITY DEVELOPMENT

"Experts" and citizens within their own communities and beyond should address environmental challenges.[46] Participation in voluntary associations has been on the increase. Although research suggests students rarely seek out service opportunities independently.[47] What is the role of education in promoting active participation within one's community? Theorists from Tocqueville (1945) to Barber (1992) have argued that the health of nations depend upon educating citizens in "schools of democracy", such as voluntary associations.[48] Below is a discussion on the development of community and citizenship participation, which are important elements in environmental education.

A task for educators is how to promote citizenship development among students and at the same time address the myriad problems that continue to challenge society.[49] Astin and Sax "suggest" one answer to promote citizenship is using the simple but extremely powerful pedagogical tool known as service learning.[50]

One advantage of service learning is its ability to engage students in the learning process. In an earlier study on the traditional information-dissemination model, it was found that students are not attending to the lectures 40% of the time, and by the last ten minutes of a lecture, students are only retaining 20% of the information.[51] [52] In contrast, there are numerous faculty testimonials about the difference service learning has made in a students' drive to learn.[53] [54] [55] [56] [57] [58] [59] However, "developmental factors need to be considered to augment comprehension."[60]

Writing on learning, Kant explained, "An individual's experience plays a key role in the formation of knowledge and new experiences enhance the knowledge base. Transformation occurs through the creation of new understandings." [61] Each new construction will depend on one's cognitive abilities and developmental levels. Assessing their competence, emotional levels, and developmental stages could enhance a service learning experience. Brook notes, "to maximize the likelihood that students will engage in the construction of meaning in new environments, professors must interpret student responses in developmental terms." [62] A resource to assess these levels is Chickering's developmental model. Chickering conducted a study on the developmental level of incoming college students and their progression through graduation.[63] Studies on service learning show that "service givers often become interested in speculating and personally inquiring about the human antecedents and consequences of psychological health conditions and how living systems of a geological, geographical and economic nature. Questions of personal identity, ones own function as stimulus and response to others and the environment, and moral and ethical issues in the introduction of change are bound to arise."[64]

In service learning, students responsible for choosing and managing their own projects develop competence, purpose and integrity, establish identity, and gain autonomy toward interdependence. Finally, it is important to remember that student interactions in a community environment depend upon their cognitive abilities to accommodate a new setting in which their tolerance for differences may be challenged.[65]

STUDENT OUTCOMES

According to Miller, early studies on student outcomes revealed that students almost universally value service learning experiences;[66] moreover: involvement in community service learning enhances self-esteem;[67] improves participant knowledge in the areas most directly related to the field experience;[68] [69] improves the integration of theory and practice;[70] and is most effective when students participate in regular discussion groups oriented to helping students reflect on and analyze their experiences.[71]

Eyler and Giles conducted a historical perspective on national studies conducted to address the impacts of service learning on students. The report summarized numerous studies that show that service learning has a small but positive effect on personal skills such as efficacy, reduced stereotyping and social responsibilities. These studies also indicate that programs with more opportunity for reflection, substantive links between coursework and service, and ethic and cultural diversity have a stronger impact.[72]

In an earlier study, Eyler (1996) found that students' service learning experience increased their belief that people can make a difference and that they should be involved in community service, particularly leadership and political forums.

A youth development study, as well as the work of Astin, Sax and Avalos, in higher education, shows that volunteer service leads to subsequent community involvement.[73] [74] Benson (1993) reports that youth who help others as little as one hour per week have half the incidence of negative behaviors such as skipping school, vandalism, and frequent use of alcohol and other drugs. Gray's study examined three areas of student development: civic responsibility, academic development, and life skills. Results were based on freshman surveys and follow-up data collected from 3,450 students (2,309 service participants and 111 non-participants) at 42 institutions with Learn and Serve America Higher Education (LSAHE) programs. The study showed that service learning favorably influenced 35 measurable student outcomes. In civic responsibility, more than twice as many service participants (60% versus 28%) reported a stronger commitment to their community. The authors state that their analyses revealed significant positive effects in academic development.[75]

Sugar and Livosky questioned the positive increases in academic areas by raising the issue of extra-credit that is often given for participating in service learning projects.[76] Furthermore, evidence suggests that higher achieving students more often select community service learning experiences than lower achieving students. Notwithstanding the criticism about the type of students that choose service learning, the participants showed a positive change in life skills, with the largest differences occurring on understanding community problems, knowledge of different races/cultures, acceptance of different races/cultures, and interpersonal skills.[77]

Other smaller studies provide evidence that service learning has a positive impact such as the complexity of problem analysis, identification of locus of problem or solution, use of information to support arguments, creation of practical strategies for community action, cognitive moral development, and critical thinking.[78] [79] [80] [81] Understanding complexity leads to environmental ethics that include an appreciation of the unique purpose of every creature.

According to recent studies, educators are preparing students for an increasingly turbulent society, which will require adaptability, sophisticated knowledge, lifelong learning, and problem solving skills.[82] [83] The traditional intellectual tasks of the classroom offer a poor match with the kinds of learning that students must practice in the workplace and the community.[84] However, these descriptions of needed learning characteristics are remarkably similar to the qualities that have long been identified with service learning.[85] Astin and Sax extensive research revealed significant differences favoring service participants including understanding of the nation's social problems, ability to work cooperatively, conflict resolution skills, and ability to think critically.[86]

According to Finger and Princen how students view and behave in a diverse society is important because different societies are now interconnected through media and trade. As the economy, environment and information become more linked globally, education pedagogy should include a progressive contextualization that requires higher order critical thinking.[87] Finger and Princen emphasize, "Humankind has never before faced environmental problems that are at once biophysical and social, and that have global dimensions".

The ability to function in a more complex society requires critical thinking abilities more advanced than those typically attained by American college students, and researchers believe the challenges and support provided in service learning programs may facilitate development of critical thinking skills.[88]

Hence, in a turbulent society students need to be versed in problem solving. Thus, service learning has the potential to aid communities in solving existing problems. At the same time, service learning students gain a better perspective of their communities' challenges.

INTERDISCIPLINARY APPROACHES

Literature confirms that service learning programs promote learning about the larger social issues behind each particular project. Service learning includes "a deeper understanding of the historical, sociological, cultural, economic and political contexts of the needs or issues being addressed."[89] However, Zlotkowski notes that reliable data on service learning in the natural sciences is scarce. However, service learning has potential for the sciences since service projects often require the skills and knowledge based in many disciplines.[90]

Pioneers in ecoliteracy promote the system's thinking approach that service learning offers.[91] Orr contends, "The response of colleges and universities is lethargic and is not what one might expect of institutions dedicated to having advancing knowledge". Table 1 outlines school reform recommendations that promote similar principles such as integrated curriculum, project-based learning, and authentic assessment.

Table 1 -Systems Instruction[92]

Systemic Understanding	Systemic School Reform
Parts ~ Whole	Single subject matter integrated curriculum Individual class periods – block scheduling
Content Process	Mandates participatory, consensual decision-making
Hierarchies Networks	Principal-driven shared leadership District decisions site based management "expert" professional development informal networks
Absolute Knowledge~ Contextual Knowledge	Teacher as expert teacher as facilitator Prescriptive curriculum-project-based learning
Quantity Quality	Standardize testing authentic assessment Measuring mapping

CRITICAL THINKING

Other smaller studies cited on critical thinking provide evidence that service learning has a positive impact including the ability to deal with the complexity of problem analysis; identification of locus of problem or solution; use of information to support arguments; creation of practical strategies for community action; cognitive moral development; and critical thinking.[93] These studies examined cognitive outcomes such as problem- solving, learning transfer and cognitive moral development. Boss notes that students' moral development is impacted based on the complexity of thinking about social issues.[94] Mendel-Reyes notes, "As pedagogy for critical thinking, service learning provides opportunities for problem-posing; gathering evidence and analyzing it; and formulating, carrying out, and evaluation plans of action."[95]

SERVICE LEARNING AND ENVIRONMENTAL SCIENCE

There has been a recent surge of research about the impacts of service learning on students. The literature scans different academic disciplines and general life skills. The current literature explores the impacts of service learning in the fields of civic participation,[96] [97] K-12 education,[98] teacher education,[99] psychology,[100] philosophy,[101] peace studies,[102] medical,[103] [104] [105] gerontology,[106] cultural issues,[107] international affairs[108] and English.[109] While each discipline has a unique set of traditions and assumptions that may be enhanced by service learning, some disciplines seem to lend themselves more readily to service learning initiatives.[110] Reliable data on the efficacy of service-learning in the natural sciences is rare.

Other studies have been conducted on the process of service learning examining the relationship between experiential learning and service learning,[111] how to assess the impacts of service learning through qualitative and quantitative research[112]; and the institution and faculty affects.[113] However, there is limited research on the potential of service learning in the field of environmental science or of differences in impacts based on gender.[114]

The proficiencies of adaptability, problem-solving and critical thinking are key competencies for environmental students. Adaptability is important because often strategies used in environmental planning and restoration need to be altered based on uncertainty involved in the management of complex ecosystems. The ambiguity and complexity of resource management and environmental policy requires critical thinking and problem solving skills. According to Mordock and Kransy, "evidence indicates that conducting real-life research enhances students' critical-thinking and problem-solving skills, feelings of empowerment, and understanding of science content."[115] [116]

Eyler, a lead researcher shows that students' service learning experience-increased students' belief that people can make a difference and that they should be involved in community service and particularly in leadership and political forums.[117] These skills should be fundamentals in environmental education programs.

In two independent surveys, 89% of incoming college students identified the environment as their top social concern; 90% of high school students said they do not know enough about environmental issues; and 84% of these high school students said they would take action to improve the environment, if they had more information about what to do.[118] According to Ramaley, service learning can be viewed as pedagogy designed to enhance learning and promote responsibility.[119] In order to address complex environmental issues, students need to be versed in the various facets of these issues which extend beyond traditional environmental education and relate to such areas as the political (i.e., public deliberation), legal, and social aspects of these challenges. Students need to learn to compromise with others in the name of the common good that is often contested and tentative.[120] Environmental planning and restoration projects are generally controversial so skills in negotiation and public deliberation are essential for progress to occur.[121] Society must move beyond the "Progressive Era" theories that became epitomized "by the neutral expert who based decisions solely on empirical measurements and methods and who was supposedly in no way tainted by political ideology.[122] Ironically, these researchers and others found that the citizens input became eclipsed in decision-making and policy formulation.[123] Given the complexity of environmental decision-making, this has grave consequences. Following the Progressive Era, the maximum sustained yield emerged based on interest group needs (i.e., ranchers interests' dominate the Bureau of Land Management). Thus, administrative "experts" and interest groups dictated national environmental policies. Furthermore, resource agencies tend to promote divisiveness and polarization of interests by asserting authority rather than sharing power. In the late 1990s, a new preference emerged which embraced landscape scaled, decentralized management, and public participation based on the philosophy of sustainable ecosystem management.[124] Environmental programs that incorporate service learning can help to prepare students for the new era of sustainability.[125] Community research can be used for social transformation and environmental change. Service learning teaches participants that there are only a few definitive answers to many of the questions facing communities.[126] Therefore, consensus-building and adaptive management are methods that students need to be versed so that they can contribute to environmental stewardship.[127]

Responsible citizenship is an element that several environmental educators have identified as a key component for successful environmentalist.[128] Further research is needed that documents the relationships between citizenship and environmental activism. This link may strengthen environmental involvement, service and leadership. Environmental service could be situated in their local environments or on a broader scale.

ENVIRONMENTAL EDUCATION LIMITATIONS

Conventional environmental programs are problematic because of their emphasis on science skills. To counteract the current situation more attention needs to be given to other pragmatic skills such as negotiating and promoting collaboration among diverse groups.[129] These skills may be gained through involvement in community projects. Open dialogue is a critical element of conflict avoidance as well as of participatory environmental education.[130] As Mendel-Reyes explains, "through service learning students are challenged to listen to a range of voices and to empathize with people different from themselves."[131]

In the article, which documented the original research agenda from the Service Learning Wingspread Conference, one of the major questions that practitioners agreed should be researched is "How does service learning contribute to the development of a social ethic of caring and commitment?"[132]

The development of the political skills is a key feature of any effective environmental program.[133] An example of political skills might be for the researchers to explain to community members how samples are taken and what the standard level for contaminants are in their area so that they can assess the contamination in relation to natural background conditions. Negotiation, communication and mediation are necessary to move from environmental impasses to sustainability. The essence of good conflict, says Collin, is using it to reconcile social justice and sustainability. Social justices and science is linked at a local, national and international level.

SUSTAINABLE EDUCATION

International agencies, such as the World Conservation Union, have altered their conservation programs to encompass a broader view of the local people including their knowledge and needs. In fact, important global treaties are now endorsing community and college environmental partnerships. In the Talloires Declaration, a global initiative for environmental literacy and sustainable development, the outlining principles include recognizing the intrinsic value of faculty community involvement to promote sustainable living as part of research and teaching; ensuring the continuity of community-university projects; building community capacity through formally accredited courses; and building partnerships that do not compromise principles of ecological sustainability.[134] Environmental service learning can achieve these principles.

Service learning may hold promise for giving voice to the marginalized.[135] Orr once said that "education as a discourse and as a practice, as an institution and as an experience should listen to its own exclusions, repressions and silences." Orr suggests that educators start an environmental revolution. He states, "Patriotism, the name we give to the love of one's country, must be redefined to include those things which contribute to the real health, beauty and ecological stability of our home places and to exclude those which do not."[136]

Current environmental literature is emphasizing local ecological knowledge. Place-based ecology includes the study of human and ecological communities, history and natural sciences. The bioregionalism approach uses an emphasis on place and community as a means to construct approaches to environmental policy.[137] An example is the shift to "ecosystem management" rather than policies based on city or county ordinances. Achieving a reflective sense of place may contribute to an ethic of caring about habitats and communities.[138]

Developing a place-based ecology in order to change one's perception of global challenges is being promoted in the environmental education field. Service learning in combination with traditional environmental curricula may enhance a student's ability to meet environmental challenges by utilizing a more action-oriented holistic approach to environmental problem solving (i.e. communication and negotiation).

Service learning has been shown to help students gain a new perceptive on issues of culture, communication, and community. Students learn that they have the expertise to provide technical support.

Again, the theoretical framework for this study was the examination if the skills noted by environmental educators as important (i.e. collaborative learning, technical skills, negotiation skills) were enhanced by students' environmental service learning experience.

In regards to environmental education deficiencies, the politics of science and gender implications were explored.

NOTES

1. Saltmarsh, 1996
2. Zlotkowski, 2000
3. Howard, 1993
4. Ibid, 1993
5. Schoenfeld, 1988
6. Katz, 1985; Gardner, 1991
7. Katz, 1985
8. Greene, 1998
9. Ibid, 1998.
10. Ben-Perez, 1990
11. Brooks, 1999
12. Ibid, 1999
13. Greene, 1995
14. Holland, 1997
15. Horton and Freire, 1990
16. Palmer,1998 hook, 1999
17. Bellah, Madsen, Sullivan, Swidler & Tipton 1986
18. Kinsley, 1995
19. Hornbeck, 1989

20. Westra, 1995
21. Gibbs 1982; Shiva 1988; Bullard 1990; Hamilton 1990; Pardo 1990
22. Alston, 1990; Anderson, 1992; Beasley, 1990; Bryant & Mohai, 1992
23. Westra, 1995
24. Choucri, 1995
25. Mezirow, 1991; King, 1994; and Kegan, 1994
26. Stanton, 1999
27. Brooks, 1999
28. Noddings, 1990
29. Zlotkowski, 1998
30. Moore, 2000
31. Gadanidis, 1994
32. Dewey, 1938
33. Dewey, 1933; Kolb, 1984
34. Cooper, 1998
35. Boud, Keogh, and Walker, 1984
36. King and Kitchener, 1994
37. Silcox, 1993
38. Fosnot, 1999
39. Bruner, 1985
40. Piaget, 1971
41. Brooks, 1999
42. Brooks, 1999
43. Eyler & Giles, 1999, Gray et al., 1999
44. Kahne, 2000
45. Zlotkowski, 2000
46. Earthwatch Institute, 2004
47. Schine, 1997
48. Mendel-Reyes, 1998
49. Barber, 1994
50. Astin and Sax (1997
51. Pollio, 1984
52. McKeachie, 1986
53. Howard, 1998
54. Bringle and Hatcher, 1996
55. Hammond, 1994
56. Hesser, 1995
57. Hudson, 1996
58. Kendrick, 1996
59. Yelsma, 1994
60. McEwen, 1996
61. Jackson, 1986
62. Brook, 1999
63. Chickering, 1993
64. Menlo, 1993
65. Miller, 1994
66. Conrad & Hedin, 1992; Hamilton & Zeldin, 1987; Markus, Howard & King, 1993; McCluskey- Fawcett & Green, 1992
67. Hedin, 1989; Wilson, 1974

68. Conrad & Hedin
69. Hamilton & Zeldin
70. Markus et. al.; McCluskey-Fawcett & Green
71. Conrad & Hedin; Hamilton & Zeldin; Miller,1994
72. Eyler & Giles, 1999; Astin & Sax, 1998; Gray, Ondaatje, Geschwind, Fricker, Goldman, Kaganoff, Robyn, Sundt, Vogelsang & Klein, 1999; Melchior, 1997; Alt & Medrich, 1994; Anderson, 1999; Eyler, Giles and Gray, 1999; Mabry, 1998
73. Youniss, McLellan & Yates, 1997
74. Astin, Sax and Avalos (1999
75. Gray, 1999
76. Sugar and Livosky (1988
77. Serow & Dreyden, 1990
78. Batchelder & Root, 1994
79. Boss, 1996
80. Eyler & Giles, 1999
81. Eyler & Halteman, 1981
82. Davis, & Meyer, 1998
83. Vaill, 1997
84. Resnick, 1987
85. Astin, 1995; Barber, 1994; Ehrlich, 1997
86. Astin and Sax, 1997
87. Finger and Princen, 1994
88. Eyler & Giles, 1999; Kitchener, Lynch, Fischer & Wood, 1997
89. Kendall, 1990
90. Zlotkowski, 2000
91. Orr, 1998
92. Barlow, 2000
93. Batchelder & Root, 1994; Boss, 1996; Eyler and Giles, 1999; Halteman, 1981
94. Boss, 1994
95. Mendel-Reyes, 1998
96. Wellman, 1999; Kahne, 2000
97. Battistoni,1997; Checkoway, 2000
98. Mezzacappa, 2001, Swick, 1999
99. Anderson, 2001
100. Bringle,1998
101. Lisman, 2000
102. Crews, 1999
103. Norbeck,1998
104. Seifer, 2000
105. Barner, 2000
106. Blieszner, 2001
107. Aberle-Grasse, 2000
108. Myers-Lipton, 1996
109. Alder-Kassner, 1997; Bacon, 1999
110. Carron, 1999
111. Moore, 2000
112. Bringle, 2000
113. Holland, 2000 and Driscoll, 2000
114. Howard, 2000

115. Wright, 2005
116. Dewey, 1938; Giles, Honnet, & Migliore, 1991; Hart, 1997; Hungerford & Volk, 1990; Morrow, 1999; Pennick, 1995; Solomon, 1997; SRI, 1997
117. Eyler, 1996
118. Renew America Report, 1989
119. Ramaley, 2000
120. Mendel-Reyes, 1998
121. Conn,1990
122. Cortner & Moote, 1999
123. Ostrom,1990
124. Cortner & Moote, 1999
125. Giles, Honnet & Migliore,1991
126. Rhoads, 1997
127. Giles, Honnet & Migliore, 1991
128. Orr, 1992; Shor, 1992
129. Gayford, 2000
130. Ausburger, 1992; Brooks & Brooks, 1993
131. Mendel-Reyes, 1998
132. Howard, 2000
133. Sterling, 1996
134. Rensburg, 1998
135. Stanton, 1987
136. Orr, 1993
137. McGinnis, 1999
138. Thomashow, 2002

Chapter Three
Unheard Voices: Politics of Science

Each of us has a passion –mine is the coral reefs – the world that requires us to be aware of diversity and humbled by our need for air – when we leave the human world behind, we experience a reverence which makes us aware – an experience so close to the heart and soul that one "feels" the knowledge within themselves.

TRADITIONAL ECOLOGICAL KNOWLEDGE AND EXPERT KNOWLEDGE

The dominant paradigm of reductionism science is widely accepted in institutes of higher education. However, the relationship between power, expert knowledge, and exploitation should be further researched. White, middle-class males have produced almost all the science. For the founding fathers of modern science, the reliance on the language of gender was explicit. They sought a philosophy that could be called "masculine," that could be distinguished by its "virile" powers and its capacity to bind nature to man's service.[1]

According to Bacon, "the discipline of scientific knowledge and the mechanical inventions it produces do not merely exert a gentle guidance over nature's courses; they have the power to conquer and subdue her, to shake her to her foundations." Bacon promised to create "a blessed race of heroes and supermen" who would dominate both nature and society.[2]

Merchant in 1981 convincingly argued that:

> modern natural science, particularly mechanics and physics, are based above all on the destruction and subordination of nature as a living organism understood to be female – and that at the end of the process, nature is considered only as dead raw material, which is dissected into its smallest elements and then recombined by the great white engineer into new machines that will obey his will.

Has this patriarchal approach led to a reductionist science that is exclusionary? Dr. Berkes, a traditional ecological knowledge (TEK) researcher and writer, notes that through the incorporation of TEK and the sacred dimensions of ecology, many of the shortcomings of the contemporary Western knowledge-practice belief complex can be addressed. Berkes provides illuminating examples of projects that institute mutual respect of knowledge systems. One project between the University of New Zealand and the Rakiura Indians involves a cultural safety contract regarding the harvest sea bird, titi. This example is not reflective of the common arrangements between colleges and community members.[3] As articulated by Shiva, a nuclear physicist who left her profession to become the Director of the Research Foundation for Science, Technology and Natural Resource Technology, exclusion of other traditions of knowledge by reductionist science is threefold:

> (i) ontological, in that other properties are just not taken note of; (ii) epistemological, in that other ways of perceiving and knowing are not recognized; and (iii) sociological, in that the non-expert is deprived of the right both to access to knowledge and to judging claims made on its behalf.[4]

The male controlled trajectory of scientific research has excluded the traditional ecological knowledge of millions of women in the fields of obstetrics, agriculture, water-resources management and cultivation.[5] Women's stories and contributions are often not represented in traditional educational programs, especially in the sciences. Recent literature has recognized the power associated with acknowledging women's ecological work through stories.[6]

Women have a long history of nurturing and protecting the planet, but their achievements remain largely undocumented in written texts. Abram discloses the historical implications and reasons behind silencing women:

> The burning alive of tens of thousands of women (most of them herbalists and midwives from peasant backgrounds) as witches during the sixteenth and seventeenth centuries may usefully be understood as the attempted, and nearly successful, extermination of the last traditions rooted in the direct, participatory experience of plants, animals, and elements in order to clear the way for the dominion of alphabetic reasons over a natural world increasingly construed as a passive and mechanical set of objects.

This disconnection from the earth and later shift towards reductionism science began the path of destruction and annihilation of the ancient reciprocity that our ancestors had with the natural world.[7]

According to the United Nations report entitled "Our Common Future", women are the single most important sector of society when it comes to caring for the earth.[8] What are the effects of minimizing women's participatory connections with nature? Scholars believe the link between women and the environment stems from experiencing the effects of domination, be it of nature or themselves.[9] Other scholars such as Sturgeon argue that women's reproductive characteristics connect them to natural rhythms, both seasonal and cyclical.[10] These temporal and spatial links to their surroundings are important. Service learning holds promise to connect participants to their surroundings particularly seasonal or temporal changes by establishing baseline data. Women tend to gravitate toward community work. As community changes, this type of community work acts as a natural history mark for planners and historians. Much has been written in recent years about intergenerational equity and its impacts on environmental degradation.[11]

Grays writes that "naming realities" is essential to women because naming fosters the power to shape reality into a form that serves the interests and goals of the one doing the naming.[12] Yet, if stories are not told, then the powers of women's experiences are lost. Women's intimate knowledge of the land needs to be acknowledged inasmuch as it may hold the key to sustainable practices. As Behn appraises, "Women are not beasts of burden but goddesses of wealth since we rear cattle and produce food, performing 98% of all labor in farming and animal husbandry."[13]

The majority of service learning revolves around the basic needs of the young, impoverished, and elderly in society. These needs are generally met by the toil of women. Service learning may be the method in which women's undervalued work and academic pursuits can be combined to alleviate the incredible schedules that women encounter on a daily basis. Recent studies show that:

> Raising kids, commuting, and receiving lower wages are handicaps to academic work which affect working women and minorities disproportionately because working class mothers are expected to take care of children, keep house, and earn money while they go to college and because minorities are paid less than whites in the job market and have twice the unemployment rate despite the major advances in education achievement they have attained as a group in the past decades.[14]

If academic institutions legitimize the work of women, society may begin to recognize their important contributions. Women choose service learning because it affords them the opportunity to combine family duties with their academic careers in various settings, such as after-school tutoring projects. Through these projects, women have gained self-respect as well as the respect of their children and community members.[15] In 22 of 39 jurisdictions that listed community environmental groups, more than 50% had women leaders.[16] Concurrently, the community is served by their work.

The myth that women have a "natural" capacity and desire to care has been challenged.[17] Rather than such an essentialist explanation, the reason may lie in socialization. Increasing numbers of educators argue that service demands a set of skills and an approach to social issues that must be intentionally instilled as part of the undergraduate experience.[18] Recent studies show that when service is optional, it is more common for girls to volunteer than boys. However, when boys share the experience, they too learn that caring functions belong to both sexes.[19] By incorporating service-learning into the sciences, women's accomplishments can be acknowledged and more males may learn caring functions.

ECO-EXCLUSIONS

It is interesting that many female environmentalists are never recognized in our textbooks or classroom discussions. One example is Devi, the inspiration for the Chipko movement. Devi was an adherent of the Bishbios religious sect, which held trees sacred. In 1798, the Maharajah decided to build himself a new palace and dispatched a crew to chop wood to fire his lime kiln. As the story goes, Devi begged the crew to spare the forest. As she clung to the tree, saying, "A chopped head is cheaper than a felled tree," the axe came down. After she fell to the ground, her three daughters each took her place defending the trees. All were killed. Later, people from 49 surrounding communities responded to the villagers' call for help. By the day's end, over 350 people had been slaughtered. The Maharajah then abandoned his plans and promised the villagers that their wood would not be cut.[20]

Today, the Chipko movement is comprised of a group of Indian women dedicated to saving forests. The movement was prompted by a decision to replace the banj forest with commercial trees and pine. This decision was made by international aid organizations without the input of native women. Men from the village were asked for input and reaped the financial benefits. However, the cash economy created destitution and drunkenness among men. For the women, the drunkenness resulted in violence and hunger for their children and themselves.[21]

The Chipko women, who traditionally walk long distances for fuel, fodder, and water, successfully resisted the new forest policy that was imposed upon them. In 1978, the Right to Livelihood Foundation (a.k.a. Alternative Nobel Prize) recognized the work of the Chipko women with an award that stated: "The Chipko movement is the result of hundreds of decentralized and locally autonomous initiatives. Its leaders are primarily women, acting to protect their means of subsidence and their communities."[22] This is but one example of important environmental work accomplished by women around the world and a prime illustration of how interests, empathy, and community participation can be linked to enrich the lives of women, children, and the environment. Although not labeled specifically "service learning," the premise of the Chipko movement is the same: aiding the community. Disregarding female ecological knowledge has led to destructive practices and the emergence of ecofeminism.[23]

Ecofeminism is a response to concerns of exploitation, both of nature and of women. It speaks to the dominant antagonistic and hierarchically world-view that Hegel and Marx discuss in their dialectics. It is important to examine the links and devaluation of nature's services and its servicers, many of whom are women. Nature's cycles are often manipulated to maximize quick profits. Thus, the global market supercedes other considerations such as long-term environmental impacts and the local impacts on agricultural communities, especially women and children. For example, the Federal Insecticide, Rodenticide and Fungicide Act allows for the export of insecticides and agricultural chemicals that are banned in the U.S. These chemicals are used to hasten the growth of crops. Shiva provides a poignant example of how the chemical companies influence local food cycles:

> Consider the crop *"bathua"* in India. Bathua is a green, leafy vegetable with a very high nutritional value that grows as an associate of wheat. Therefore, when Indian women harvest bathua in the wheat field, they do not merely contribute to the productivity of wheat; they actually harvest a rich source of nutrition for their families. Recently, however, bathua has been declared a "weed" by international organizations and Indian authorities. Consequently, bathua is now treated with herbicides. The food cycle is broken; women and children are deprived of a free source of nutrition.

Shiva does not mention the potential effects of these chemicals on the health of residents. Those effects might not become apparent for years.

Chemical companies are beginning to dominate university programs, especially at environmental and agricultural colleges. Consequently, chemical companies can influence what is considered "expert knowledge" in institutions of higher education. Students may not question their programs, which increasingly are subsidized and controlled by multi-national companies.[24] In addition to public colleges, even private institutions are soliciting "revolution research" in order to receive funding that helps them to compete as an educational business. Companies are courting scientists and entire departments with multi-million dollar contracts.[25] How can this be combated in terms of college participation?

EQUALITY AND EDUCATION

Community colleges provide high-quality access to higher education - a democratization of learning that has produced nearly 1,200 community and technical colleges nationwide. During the 100 years since the first junior college was founded, the broader community college mission has expanded from university preparation to include technical and vocational education, basic skills training, and, most recently, workforce development. Community colleges often provide retraining for older students, allowing them to make a fresh start, and career training for women who have chosen to reenter the public workforce after caring for their children and families.[26] However, at community colleges, such as Diablo Valley College, which is ranked as the third highest transfer institute of 108 community colleges in California – a small percentage of science students are minorities. At DVC, 20% of science students are African American and 10 % are Hispanic. The science field leads to higher paying jobs. By not serving minorities, community colleges may be contributing to their economic disadvantage.

GOWN TO TOWN CONNECTION

The advantages of community colleges include open access, responsiveness to community needs, and student-centeredness.[27] However, the way a college interprets "education" and "service" affects how it understands its mission.[28] The goals of responsiveness to community needs and student- centeredness are often cited in college mission statements. Yet, the practical applications of these goals remain largely unmet if one compares mission statements and course offerings.[29]

U.S. colleges and universities often publicize their commitment to the community by participating in programs such as Campus Compact, an organization aimed at improving community and college relations through service learning. Campus Compact was established by Presidents Healy (Georgetown, DC), Kennedy (Stanford), and Swearer (Brown) in response to a sense of disengagement by students from public and community service in the early 1980s.[30] However, community outreach, at least through service learning, is not highly institutionalized in terms of funding and institutional priorities.[31] It seems ironic given the fact that these same institutions depend on financial subsidies afforded to them by their communities. Only a few major universities--including Providence College, Portland State University and Rutgers University - require students to participate in community projects.[32]

A college can serve community needs by offering office space, computer resources, meeting rooms, and technical expertise. However, what is often missing from these partnerships is an institution's most valuable resource - its students. Students add the element of diversity. They also offer enthusiasm and passion. For example, the history student who is passionate about uncovering artifacts from another era could be a valuable asset to a community conducting a historical or cultural study.

As Reardon, a service learning pioneer, explains, it is important that a campus be directly involved in enabling community leaders to strengthen their communities. Programs should be structured in a way that the community's agenda continually drives what the students and faculty do and learn through the partnership.[33]

The passage of the National and Community Service Trust Act of 1993 created a wide range of opportunities for citizens to address issues such as education, human services, public safety, and the environment. The Act was intended to empower citizens to rebuild their communities. It is through such national programs that citizens can begin to consider their roles in community planning and decision-making.[34] Historically, the perspective of the community in service learning programs has been neglected. Checkoway posits strategies to incorporate the service agendas of local agencies with the learning agendas of higher education institutes. Checkoway asserts that "working together to establish principles that are locally relevant will provide a forum for issues to be raised, discussed and defined before problems begin." Honnet and Poulsen elaborate on other community empowerment strategies in a Wingspread Special Report that was based on extensive deliberation by more than 70 organizations involved in service learning:

1. Provide structured opportunities for people to reflect critically on their service experience;
2. Allow for those with needs to define those needs;
3. Clarify the responsibilities of each person and organization involved; and
4. Include training, supervision, monitoring support, recognition, and evaluation to meet service and learning goals.[35]

Based on the expertise of the Wingspread participants: "the needs of community members to express their own needs in their own communication styles will ground the student participants and addresses some of the service receivers' deepest needs of all – to be heard and affirmed. This contributes to the process of self-empowerment in meeting those needs."[36]

One advantage of service learning is that it can address the otherwise unmet needs of a community. Environmental degradation certainly falls within that category, especially in poorer areas of inner cities. For example, air pollution levels in the Washington, DC metropolitan area are higher in poorer areas of the city, where most African-Americans live. A similar situation exists in New York, Chicago, Denver, Los Angeles, and San Francisco.[37] Grassroots leaders have emerged from groups of concerned citizens (many of them women) who see their families, homes and communities threatened by some type of polluting industry or government policy.[38] Green in 1998 articulated the importance of student connections to their community in a lecture to future teachers:

> When people lack attachments, when there is no possibility of coming together in a plurality or a community, when they have not tapped their imaginations, they will be unlikely to think of breaking through the structures of their world and creating something new.

Traditionally, colleges have been slow to formally and systematically aid community groups. Rarely, if ever, is service learning given more than a token acknowledgement with certain exceptions such as Dartmouth College, Colby College, Bates College, Brown University, Middlebury College, University of Vermont, University of Michigan and John Carroll University.[39] This study focuses on three projects. Several pilot projects were also conducted. The next section explains the design, methodology, site selection, and data analysis procedures that were used in the case studies as well as a brief explanation of my pilot study.

NOTES

---

1. Keller, 1985
2. Spedding, 1963; Keller, 1985
3. Berkes, 1999
4. Shiva, 1997
5. Ibid
6. Troy, 1996
7. Abram, 1996
8. Caldicott, 1992
9. Merchant, 1981; Shiva, 1997; Hamilton, 1990; Adair, 1990
10. Sturgeon, 1997
11. Weiss, 1989
12. Troy, 1996
13. Behn, 1980
14. Pincus, 1995
15. Russo, 2002
16. Taylor, 1997
17. Abel, 1990; Hooyman and Gonyea, 1995
18. Wutzdorff, 1997
19. Pardo, 1997
20. Brenton, 1998
21. Shiva, 1997
22. Right to Livelihood Foundation, 1987
23. Warren, 1997
24. Kenney, 1986
25. Ibid
26. Bok, 1982
27. Boyer, 1994
28. Stanton, 1999
29. Ward, 1996
30. Stanton, 1999
31. Wutzdorff and Giles, 1997
32. Reardon, 1998
33. Reardon, 1997
34. Checkoway, 1994
35. Honnet and Poulsen, 1989
36. Tice, 1994
37. McCaull, 1976; Gelobter, 1988
38. Gibbs, 1982; Shiva, 1988, Bullard, 1990, Hamilton, 1990; Pardo, 1990
39. Ward, 1999

Chapter Four
Western State College and Rutgers University Case Study

When the waters again run clear and their life is restored – we might see ourselves reflected whole.

David Orr

EMPOWERMENT IMPLICATIONS

My doctoral research was conducted in three main locations, Leadville, Colorado, New Brunswick, New Jersey and Gunnison, Colorado. The institutions involved included Colorado Mountain College (e.g., pilot) Rutgers University and Western State College. Rutgers was chosen because of their institutional leadership in the area of service learning, Western State was in the infancy stage of developing service learning but key faculty namely Dr Jessica Young and George Silby held a deep commitment toward environmental sustainability in the West. A sense of empowerment is important to promote engagement within a community. A leading activist, Bryant, promotes the use of action research for facilitating social and environmental change. He encourages students and faculty "to gather data to expose environmental dangers so that those community members affected can make informed decisions about changing their own communities for environmental betterment."[1] Through local action projects, students and citizens play a more significant role in governmental and development projects that affect their lives. Thus, the work focuses on the integration of local and academic knowledge through collaborative efforts among students, educators and citizens. For example, the bio monitoring of a stream on the Clark property in Gunnison, Colorado was conducted by select community members, student participants, faculty, United States Geological Society professionals, and me. The field observation was helpful for understanding the lack of knowledge among community members, faculty and students regarding the legal requirements for conducting sampling and transporting samples for analysis.

My work incorporated action research, which is meant to have the participants own activities inform their ongoing inquiry. Action research gives credence to development of powers of reflective thought and action. "No action without research, no research without action." Furthermore, Taggart notes that authentic participation in research means sharing in the way research is conceptualized, practiced and brought to bear on the life world.[2] Action research draws on the precepts of emancipation. Friere promotes emancipation as central to the educational process.[3] Other studies recognize that sustainable empowerment and development must begin from the concerns of margined.[4] The "marginalized" in the Western State study is the flora and fauna in a rural Colorado community. As Dr. Seuss once stated, "I speak for the trees for they have no tongues." The participants included college participants, community members, city planners and state officials. The research of the creation and monitoring of artificial wetlands combined monitoring skills and collaborative approaches to community planning and restoration. These projects and participants adhere to recommendations of service learning experts. Namely that, service programs should engage students in responsible and challenging actions for the common good. This involvement should empower both those who are being served (i.e., community members) and those who serve so that service does not become one-way.[5] The scientific research such as the nesting habits in deadfall informed the city planners about land-use regulations' impact on wildlife, thus affecting tourist trade. These initial projects may have lead to a collective partnership, which promotes change and improvements in environmental planning and restoration. The project educated students about the social and legal aspects of their work. The grounded theory research appraised the educational impact of environmental service learning including cognitive, affective and psychomotor skills. The cognitive skills centered on ecological and scientific aptitude while the affective and psychomotor impacts focused on aesthetic and moralistic conduct (See Figure 1). Again, this connection is important because as Palmer states, our educational institutions must not be divided from society and student's inner personal values and motivation.

The cognitive skills included monitoring, observation, systematic thinking and development of local ecological knowledge. The affective skills consisted of examining their attitudes and beliefs over a five-month period. Psychomotor measures tested changes in their behavior. The cognitive, affective and psychomotor measures were addressed through a series of questions such as:

1. Can students benefit by working collaboratively on local environmental issues?
2. How can service learning enhance environmental education specifically the exclusive emphasis that environmental education places on science?
3. How can service learning enhance subject matter learning?
4. Does service learning contribute to the development of an environmental ethic of caring?
5. Are there gender differences in the impacts that students express from this experience?

Pre-Experience Perceptions

In their initial interviews, students listed "cooperation" and "responsibility" as lead terms to describe service learning (See Figure 4). Students described community service as a “punishment” and “work without any monetary compensation”. Their expectations were that they would gain “hands-on” experience but that the experience would entail a substantial time commitment. Questions were about general environmental issues as well as project specific impacts. Ninety percent of the participants noted that overpopulation was the most pressing environmental issue. The remaining participants cited lack of education and public complacency as major environmental problems. Participants thought of themselves as lacking in legal knowledge, lacking in any meaningful connection with the community and not able to understand the social aspects of environmental work. At the same time, students placed a high value on for their responsibility to future generations and expressed the belief that one person can make a difference.

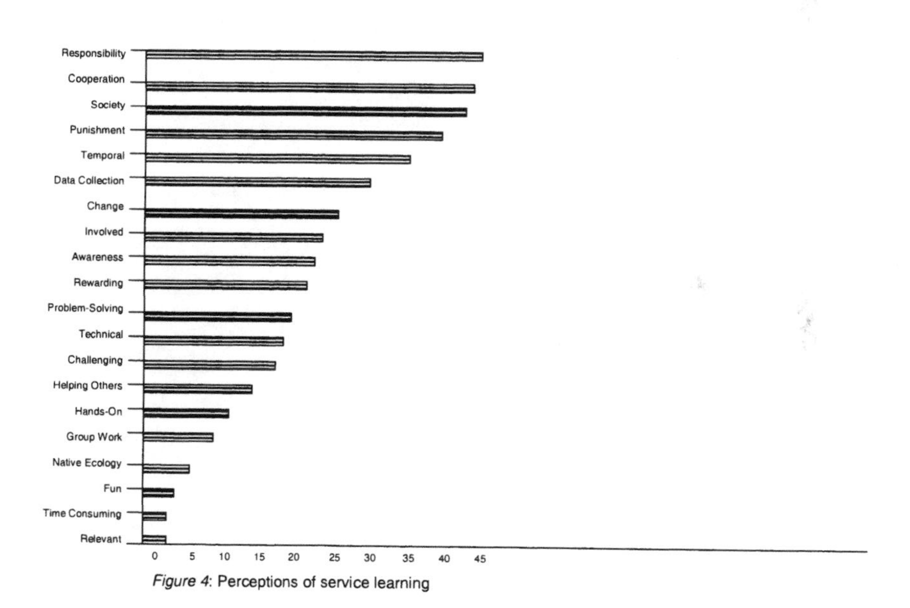

Figure 4: Perceptions of service learning

As indicated in Figure 4, relevance was the least anticipated outcome. However, in the final interviews, participants rated relevance as one of the highest impacts. In terms of cognitive outcomes, participants ranked data collection and problem-solving skills as the most anticipated outcome.

MID-EXPERIENCE FINDINGS

During the eighth week of the study, a mid-project survey explored how the service- learning course compared with traditional courses. Participants reported general reflections on the experience in terms of what they learned, how they enjoyed the experience and whether the experience changed their worldviews or their views of themselves. They reflected on how the actual experience differed from their preconceptions. Lastly, participants commented on the salient aspects of the experience. The validity of qualitative research is based on its "grounded" connection to the data.[6] Thus, the direct statements (i.e., data) of some of the participants are included.

In general, participants remarked that their experience was not only more enjoyable and rewarding than a traditional class but that they learned dramatically more. For example, students stated:

1) "I learned ten times what I would have in a traditional classroom."

2) "Most other classes don't even compare in regards to learning."

The follow-up question asked students to give specific examples of what they learned. Some of the responses included the following: "group work, professional standards, work ethics, field techniques; discipline; patience; biodiversity and native ecology."

In response to the questions regarding whether the experience altered the students' worldviews or sense of themselves, the majority of students noted that the experience taught them the importance of volunteering and acting as agents of change. (See Figure 5.)

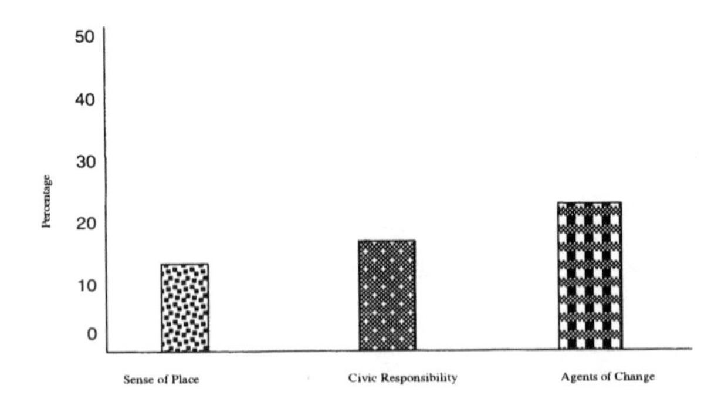

Figure 5: Behavioral and Local Ecological Knowledge Impacts

They added that they perceived the experience as "valuable, fun and positive." Another finding was that the participants gained a deeper understanding of their local environment as well as the process of group dynamics. They noted that it was exciting to know that "their data was actually used and valuable" in terms of formulating and evaluating land-use plans. A vast majority of the participants stated that the project helped to improve their skills and perceptions on data-collection, problem solving, and critical thinking and negotiation skills. After the experience participants noted that their scientific skills particularly sampling improved.

Similarly, participants stated that their critical thinking and negotiating skills improved by 90% and 91% respectively.

According to the participants, the sentient aspects included the temporal dimension, technical skills (i.e. monitoring, sampling), practical skills (i.e. speaking and organizing meetings), and the relevance of quality data collection techniques.

POST EXPERIENCE FINDINGS

Ten major outcomes emerged from the participants' responses about the values of environmental service learning: societal context, satisfaction, cooperation, civic responsibility, relevance, their role as agents of change, broader appreciation of the discipline, academic integrity. The outcomes are based on the following descriptive words:

1. Academic Integrity: "time-consuming", "challenging", "discipline", "informative", "difficult", "theoretical", "career-related"
2. Agents of Change: "change", "difference", "transform", "temporal"
3. Broader Appreciation of the Discipline: "technical skills (tools, equipment), "problem-solving", "data collection", "monitoring", "know"
4. Civic Responsibility: "involved", "responsibility"
5. Confidence: "ability", "pride", "assurance", "one's capacity", "capability"
6. Cooperation: "group work", "helping others", "cooperation", "people", "relationships", "with others", "classmates", "interactions", "comfortable", "community"
7. Sense of Place: "native ecology", "grasses thistle", "wetlands", "perceive environment", "awareness"
8. Societal Context: "punishment", "society", "community", "people", "community organizers", "giving", "relationships", "communities", "residents", "representatives", "interactions", "connections"
9. Satisfaction: "fun", "good", "hands-on", "active", "rewarding", "better classroom", "wonderful", "valuable"

10. Relevance: "real", "realistic", "relevant", "applicable", "germane", "related"

After the experience, participants ranked responsibility, cooperation and societal connections as the most important impacts (See Figure 6).

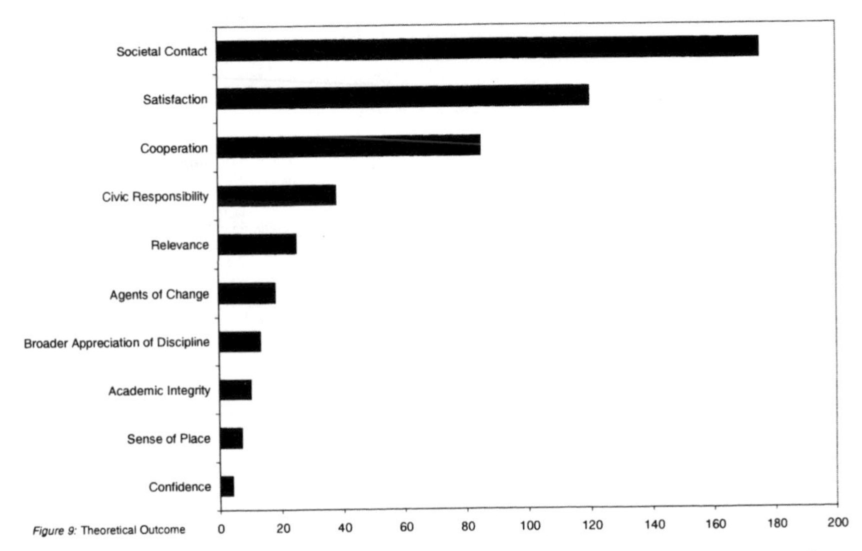

Figure 9: Theoretical Outcome

Students consistently remarked how service learning impressed upon them the significance of cooperation. Respondents said, "It was a positive experience working with all different types of people; this experience has taught me much more about equal and mutual efforts with a variety of different people and it gives people an opportunity to work together, and that is what is missing from traditional education."

The ability and preference to work with others is especially important for women in the science field. According to an earlier study, women may experience a sense of personal failure at the college level because of its emphasis on autonomy, and their male colleagues may view them as less competent because of their desire for connection.[7] This research concurs with that study.

SUMMARY& POST SCRIPT

To summarize, participants ranked placing scientific work within a societal context and satisfaction as the most significant outcome of the experience. Throughout the investigation, outcomes were developed based on the students' words. The students' language is the essence of the study. The framework of the study is post-positivism. This research centers on understanding the phenomena from a person's individual perspective. Educators and researchers refer to this approach as the "emic perspective" inasmuch as the investigation focuses on the viewpoint of the participant rather than the researcher.[8]

Again, the outcomes relate to impacts on collaborative learning, subject matter enhancement and changes in ethics.

In rating their service learning experience, 30% students placed an emphasis on the sense of satisfaction they gained from participating in a service-learning project. The category of satisfaction is noteworthy given it was rated as the second highest outcome. Many students identified apathy in their pre-experience interviews as the most pressing environmental issue. Therefore, if environmental service learning is perceived as "wonderful", "rewarding" and "fun", fewer citizens may be apathetic toward environmental protection planning and preservation projects. In addition, acknowledgement of the affective dimension of this finding is important because emotions often serve as a source of insight, particularly for female scholars.[9]

According to narrative analysis, the importance of placing science within a societal context and learning to cooperate was a strong outcome. 78% students commented that learning to cooperate was key to their experience. Humans act toward physical objects and other beings (i.e., flora and fauna) in their environment based on the meanings these things have for them; these meanings derive from the social interaction and these meanings are established and modified through an interpretive process. Therefore, if students are active in the environment, the meaning of the environment can be established and modified. Thus, service learning creates a learning space that connects its participants literally and figuratively to their environment. Students claimed the environmental experience was valuable and strengthened their connection to the environment. According to previous research a holistic understanding (i.e., societal context) aids interpretation of data.[10] This is an essentially important skill in the field of environmental science.

Participants believed the service learning experience had academic integrity (i.e., challenging and informative). They noted that it gave them a broader appreciation of the technical skills (i.e., monitoring, data collection) required in the field.

The positive rating for academic integrity is interesting because science educators tend to think that service learning lowers the academic standards. Yet, science related service learning teaches students how to apply the scientific method and helps them appreciate the unexpected problems and stresses associated with doing original research.

In a nationwide survey of over 4,000 secondary schools, students reported learning "much more" in their service programs than in their regular classes.[11] Furthermore, in a more comprehensive nationwide study of college students, results indicate that participating in service during undergraduate years substantially enhances a student's academic development, persistence in college, and interest in graduate study.[12]

The 553 citations of the key codes answered the two leading cognitive questions namely that service learning improved collaborative learning and subject matter enhancement. Participants ranked an appreciation of how their scientific work fit within a societal context as an important outcome.

Most respondents cited the positive impacts of cooperative learning and improvements in their confidence levels.

In terms of the affective and psychomotor impacts, there was an increase in students' discernment of their personal responsibility, their scientific skills ratings, and enrichment in their negotiation skills and perceptions that individuals can make a difference.

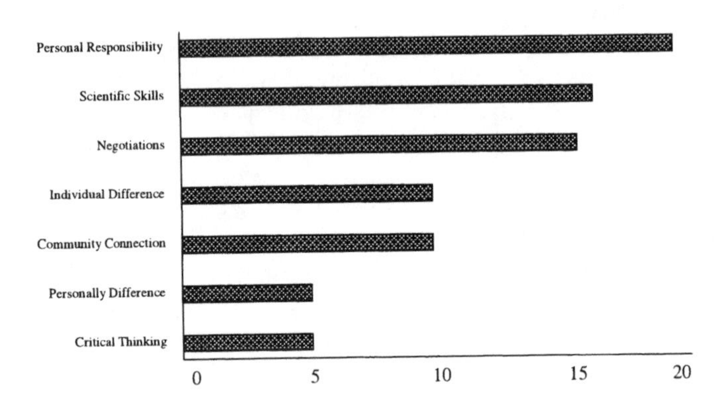

Figure 7: Transformation Assessment

Figure 7 results illustrate responses to open-ended questions asked during the three interviews. The number of times participants used the terms produced the rankings. During the interviews the participants were also asked to rate their perception of their skills attitudes and behavior prior to and after the experience to determine changes in their perceptions. Students thought that individuals could make a difference in solving environmental problems after their service learning experience. A few participants noted a decline in the idea that individuals could make a difference – perhaps because they learned the complexity of research and political decisions.

These changes are again, based on students' perceptions. Ward notes:

> Most classroom higher education is a solitary experience for students, with the emphasis on individual performance in response to exams and assignments. It provides no experience in-group problem solving. In contrast, environmental service learning commonly relies, like most professional environmental work, on shared expertise of groups to provide a broader range of skills and knowledge.[13]

An improvement was noted in communication skills and the perception that they personally could make a difference.

In summation, the respondents clearly categorized differences in the areas of enhancing a student's sense of personal responsibility and improvement in scientific and negotiation skills.

Additional post-experience questions measured changes in student perceptions regarding their role in society, ability to get involved, awareness of issues and social activism. All participants noted that service learning was a valuable experience. The particular question asked students to evaluate if the experience was valuable An overwhelming percentage of the participants responded that they either strongly agreed or agreed.

At least one environmental professor stresses that "the process of building a constituency of young people who understand environmental problems at the community level may turn out to be more important than the scientific advancements that we also need to reverse our present and unsustainable course."[14] A majority of the students noted that service learning improved their awareness of environmental issues and more importantly, the perception that they should become involved in community projects.

Ninety-five percent of participants stated that their environmental awareness improved based on the experience as well as their beliefs about individual efforts and societal efforts to improve environmental issues .

During the post-experience interview, 37 % of students noted enhancement of subject matter as the following as the most rewarding part of the experience, followed by 32% rating helping others; 26% research skill improvements and 5% stressing collaboration. Specific language is as follows:

1) gaining knowledge;
2) helping others;
3) producing valuable data for USGS;
4) conducting field work;
5) helping community;
6) observing a difference in the environment;
7) knowing that the project data was actually going to be used;
8) seeing final product;

9) learning and helping;
10) motivation;
11) paying attention in the field;
12) experience;
13) connecting with the community;
14) seeing new things;
15) working with others;
16) knowing helped community;
17) effecting change;
18) getting residents involved in local planning;
19) knowing what I did made a difference.

Fifty percent of students identified aspects of academic integrity as the most challenging aspects of service learning; followed by a 25% rating for systems thinking and 25% rating for communication or negotiation. Specific comments included:

1) time consuming;
2) reflecting and documenting the experience in the interviews;
3) keeping a good journal;
4) trying to comprehend so much;
5) starting field set-ups;
6) some people were hard to work with;
7) comprehension of benefits for future generations;
8) understanding the scope of the work;
9) preparation;
10) physical work;
11) coordination with a group;
12) lack of leadership.

It is interesting to note that numerous theoretical codes contained dichotomies based on traditional approaches to college level environmental education and the service learning experience. For example, students commented on the cooperation vs. competition methodologies as well as broader discipline vs. compartmentalization learning.

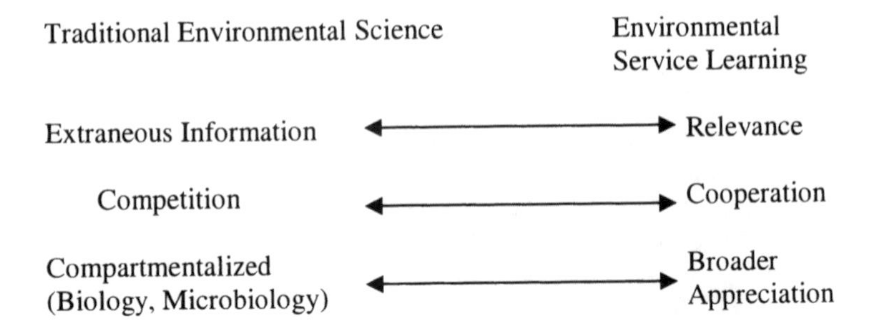

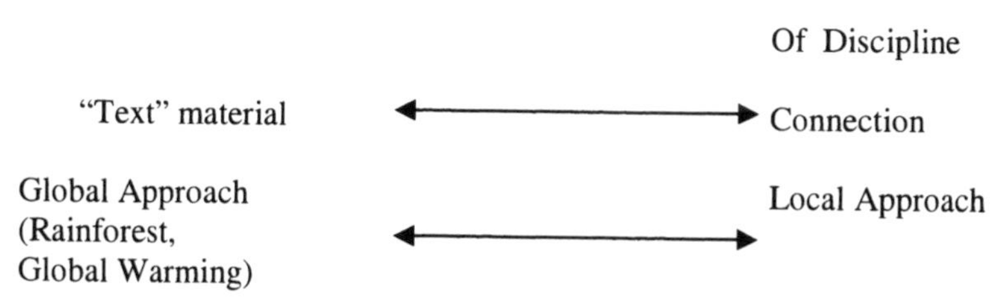

Table 2: Environmental Science Dichotomies

Environmental educators have identified dialogue as a critical element for environmental education.[15] Students rated a 5% enhancement of their communication skills after the experience. Communication is critical in the environmental debate. The views and terms used by students who might be classified as preservationists and conservationists became apparent during the interview process.

Given that the majority of respondents were environmental science students, it is surprising to note that self-ratings were low in the areas of scientific and negotiation skills. Professors at Colby College note that although students have little exposure to specific regulations and public entities, they continue working directly with state or local agencies after their service learning.[16] Only 70% rated themselves proficient in these areas prior to the experience, so service learning might conceivably enhance science and negotiation skills by working directly with professionals in the field. These areas could be addressed through curriculum revisions.

Other dichotomies found in the participants' views exposed the limitations of entrenched epistemologies. For example, several students noted that studying local issues such as bird habitats put environmental restoration into context for them. Consequently, they began to view the intrinsic value of certain species. As stated by a leading biology professor, "ecological problems are recognized through observations made in the field and therefore such observations are vital to the science. To become preoccupied with theory or with methodology to the exclusion of looking at nature is surely to lose contact with the world that we are trying to understand." [17] Interestingly, in earlier studies of environmental service learning action projects, many students changed their majors because of their experience.[18]

CONCLUSIONS

The theoretical codes that arose from the study are divided into categories related to the leading questions. The codes are categorized into such areas as collaborative learning, emphasis on science, subject matter enhancement, and ethics (See Figure 8).

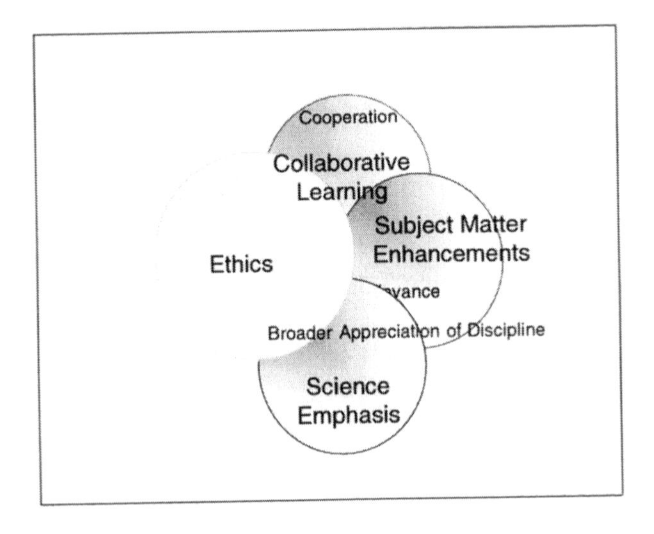

Figure 8: Summary Data

The impacts in the four general categories appear to interact in several areas. For example, participants related having a broader appreciation of the environmental science discipline that enhanced their skills and placed their learning objectives into a much broader context in terms of science and political realities. The self-ranking indicates that the experience showed impacts on collaborative learning specifically, the satisfaction and ability to cooperate among different stakeholders. For example, the bird study links to local planning issues. If the participants can illustrate a decline in bird species based on loss of habitat – it could influence land-use planning in the area. Citizens were trained in sampling and monitoring techniques at the three sites. The bird habitat study was the baseline study so further research will need to be conducted.

According to previous studies, females are more inclined to participate in collaborative projects, so environmental service learning may enhance their participation in the sciences.[19] Collaboration and systemic thinking are areas where females excel, so by offering environmental education in a different format, potential success could be realized by female students. [20] These particular skills (i.e., collaboration, community involvement) are identified as essential in environmental law and policy in order to achieve ecological sustainability.[21]

The remaining questions centered on subject matter enhancement, ethics and the emphasis on traditional science in the curriculum. The research indicates that students valued using their data within the community, followed by satisfaction, ability to cooperate, and a heightened sense of civic responsibility. These findings are consistent with earlier studies that found service learning participants exhibit a greater sense of civic responsibility after their involvement. [22] All participants agreed that the experience has changed their affective and psychomotor deportment. The civic responsibility findings have gender relevance since females tend to volunteer within their communities in addition to shouldering their academic workload. If service learning can offer opportunities in civic projects that they are currently participating in, more female students may enter and remain in the science field because their overall workload will diminish. The link to civic responsibility is important because leading environmental educators state that responsible citizenship is a key component for successful environmentalists.[23] In addition, a Wilderman case studies on power and science education states:

> Just as students feel a sense of empowerment from their ability to apply skills they have learned to help solve problems, so do community members who are trained to become their own experts. It also opens up previously closed doors for meaningful participation in the decision-making process. Empowering lay people helps depose of the "elite priesthood" of scientists, who have very little sense of the importance of public participation in solving of community problems.[24]

Related to this, earlier research indicates that students in field placements develop collegial relationships with agency personnel.[25] Developing links to state officials on a voluntary and paid basis will strengthen graduate and community member's skills; enhance student's marketability and enhance the environment.

The United Nations asserts that behavior change follows access to information and increased involvement in environmental decision-making.[26] According to Desta, well-informed citizens are more likely to hold governmental agencies accountable.[27] Education is essential in the rapidly changing environmental field.[28] Service learning can fulfill the mission of improving access to information and facilitating negotiated solutions among stakeholders.

Participants concluded that the experience provided them with a new perspective about their ability to solve community issues and the effort that is necessary for environmental protection and planning. Participants noted that the experience provided them with a new perspective on environmental issues. As discussed earlier the majority of participants also stated that their experience positively influenced their perception that both at an individual and societal level, environmental challenges can be overcome.

Students identified the differences between their traditional college-level environmental curriculum and their preference for environmental service learning experience. They noted that service learning enhanced their ability to cooperate, their comprehension of the discipline, and their local ecological knowledge. With over six billion people on the planet, it is time to reexamine the connections between humans and the environment. A more comprehensive examination of the discipline may reveal that science education reflects Western values that devalue nature.

In contrast to the perception that environmental science "waters down" science programs, students valued the experience as having more academic integrity than the traditional information-dissemination lecture model. They also valued it for improving systemic thinking. Recent studies stress that environmental education should enable students to extend their horizons by exploring political, economic, and other aspects of education rather than being concerned solely with scientific content. [29] Surveys from leading environmental science programs indicate, "Many types of alum stated that they would have liked a skills-oriented course that linked them to real–world clients solving real-world environmental problems. Palmer argues that connectedness is the principle behind good teaching. He notes that our Western commitment to thinking in polarities is a process that elevates disconnection into an intellectual virtue.[30]

In the fourth theoretical core, students indicated the experience edified their sense of civic responsibility and ability to be agents for change. In the post experience interviews, 5% of students indicated that they gained a stronger community connection. In conclusion, the findings indicate that the environmental service learning has a transformative effect particularly in areas such as societal context of their academic work, civic responsibility, satisfaction, cooperation, and relevance. Again, these are parameters identified as important by environmental educators .[31]

The fifth leading question was designed to assess gender differences in the service learning experience. Studies on the biases of standardized tests have found that "women think in complexes rather than simplexes and work collaboratively rather than individualistically". [32] Fifty-seven percent of the female participants revealed a preference for service learning and noted that it enhances collaborative and complex thinking. Thus, it is more aligned with a female learning style. Thus, female students rated their comfort level in service learning vs. traditional science classes.

Responses were:

> 1) "Yes, I felt more comfortable for many reasons. I was one of the few seniors and had lots of experience from the summer internships. The past science classes (physics, organics, biology) had all been taken in large lecture classes with a lot of cut-throat pre-med students, so I don't know anyone who could be comfortable in that situation."
> 2) "I liked this class more because it gave me a chance to understand monitoring better."
> 3) "It was better because of the hands-on and group work. It helped me apply my interests to science."
> 4) "Service learning is preferable because it helps develop practical skills, exposes students to professionals in a field setting, it positively impacts the community."

> 5) "I like both service learning and other science classes because I always learn new, cool things. I enjoyed the opportunity to work with and talk to people I would not have normally interacted with."
> 6)"No, I don't like service learning better mostly because I am right-brained."

In summary, females rated environmental service learning higher in comparative effectiveness. Earlier studies confirm that service learning projects contextualize scientific thinking thereby reducing a sense of irrelevance felt by some students especially women and minorities in traditional science. [33] Studies show that understanding and action are the cornerstones of effective environmental education. These indicators suggest that there should be a focus on student-centered approaches to environmental education, especially for female students.

Credit for service learning experiences could be accepted on graduate-level admission applications since basic standardized tests still disadvantage females.[34] These are a few small steps to challenge the racism and sexism present in our educational institutions. Some data suggests that service learning experiences increases retention and success of students of color.[35]

Service learning research and publications could also be given more recognition in graduate-level entrance requirements, since there seems to be a clash between women's preferred modes of scholarly research (i.e., collaborative learning) and publication patterns.[36] Several studies suggest that women are more likely to participate in a broad range of service learning programs. [37] At the University of Michigan, female and African-American students are regularly over-represented in service learning courses.[38] In addition, the female perspective should be incorporated within community decision-making since women's knowledge involves more than "abstractions and wives tales". [39] Curtin argues that:

> Real eco-development cannot be sustained, however, unless distinctively women's practices are granted the conceptually central places they deserve.

NOTES

---

1. Bryant, 1990
2. Taggart, 1997
3. Friere, 1970
4. Park, Brydon-Mill, Hall & Jackson, 1993
5. Kendall, 1990, Newman, 1990 and Mintz Hesser, 1996
6. Schmidt, 2000
7. Fuehrer, 1985
8. Gall, 1996

9. Cook, 1998
10. Geertz, 1973
11. Conrad and Hedin, 1992
12. Astin and Sax, 1998; Astin and Sax and Avalos, 1999
13. Ward, 1999
14. Trelstad, 1997
15. Ausburger, 1992; Brooks and Brooks, 1993
16. Firmage, 1999
17. Hairston, 1989
18. Kaufman, 1993
19. Belensky,1986
20. Ibid, 1986.
21. Rensburg, 1998
22. Astin and Sax, 1998; Giles and Eyler, 1994; Rutter and Newman, 1989
23. Orr, 1992 and Shor, 1996
24. Wolderman ., 1999
25. Pataniczek and Johnson, 1983
26. Thibodeau, 1984
27. Desta, 2000
28. Cortese. 1992
29. Gayford, 2000
30. Palmer, 1998
31. Orr, 1992
32. Belensky, 1986
33. Rosser and Kelly, 1994
34. Wenniger, 2001
35. Roose, 1997
36. Ward,1991
37. Astin & Sax, 1998
38. Chester, 2000
39. Curtin, 1997

Chapter Five: The Nature of Race

Traditionally, the backyard was a nursery and each peasant woman, the Sylviculuralist. It was significant because the humblest of species and the smallest of people could participate in it.

V. Shiva

As outlined in Chapter Four, service learning can enhance such skills as cooperation, civic responsibility, and broader appreciation of the discipline and a sense of civic responsibility. Participants noted positive cognitive, affective and conative (i.e. behavioral) influences. For example in the areas of thought, participants expressed that they developed a "sense of place", "broader appreciation of the discipline" and an "appreciation for nature." Affective changes noted were "satisfaction", "confidence" and "civic responsibility." Lastly, the conative results were participant's perceptions of themselves as "agents of change." Females cited working collaborately as a preference over traditional science instruction and both sexes noted a preference for field work vs. lectures The civic responsibility findings are consistent with much of the previous research on service learning but the work expands the literature in terms of the science emphasis and the methodology used in conducting and analysizing the research. The study differs from the majority of service learning research because the participant's words were be used to develop a grounded theory on the student's impressions of how service learning affected them. Existing literature centers on narratives and surveys of students. By using a grounded theory approach students words demonstrated the affects of service learning for enriching environmental education. In this chapter, the relationships are explored between this research and recommendations for environmental education reform. Environmental educators, writers and philosophers are proposing the incorporation of certain skills that the service learning pedagogical model can achieve.[1] These skills such as a connection with society and appreciation of nature were positive outcomes of the experience. Furthermore, current legislation initiatives are promoting these identical approaches (i.e. community partnerships). Additional research recommendations are in this chapter. These recommendations apply to environmental science professors, administrators, policy writers, lobbyists and school board members.

COLLABORATIVE LEARNING

This research shows that service- learning outcomes can achieve the skills set forth by environmental educators such as negotiation (See Figure 7 and 8). Environmental educational programs should include pragmatic skills, such as promoting collaboration among diverse groups, community empowerment, and negotiation.[2] The Earth Summit Outcomes (i.e., United Nations sponsored conference) lists partnerships and the empowerment of marginalized groups as critical elements for environmental protection.[3] Moreover, the environment is a subject that fosters a commitment to service. According to Daloz, a direct involvement in nature and the outdoors results in a larger connection to the world.[4] Collaboration encompasses skills such as communication, negotiation and understanding.

In order to sustain college-community partnerships, strong communication skills are needed. Participants initially defined their negotiation skills as limited. Through service learning, students are required to listen to a range of voices and empathize with people different from themselves.[5] Likewise, the voice of nature itself needs to be heard. As pointed out by Manes:

> Nature is silent in our culture (and in literate societies generally) in the sense that the status of being a speaking subject is jealously guarded as an exclusively human prerogative.[6]

Environmental education needs to be more inclusive and democratic.[7] We need to develop a curriculum that is more compatible with pluralistic ways of thinking about the world.[8] Based on arguments made by Dewey, the role of public education is to give everyone in society enough training so they can contribute their own view and experiences to the democratic process.[9] By actively involving students and community members and by soliciting participants viewpoints regarding optimal solutions to local problems, it is more likely that strategies will be identified that evoke broad-based support.[10] Service learning enhances these ideals.

SOCIETAL CONTEXT

Service learning holds potential for improving environmental-education programs inasmuch as students can improve their communication skills and their ability to foster change within their communities.[11] This research showed that students ranked placing their environmental science work within a societal context as the most important impact. Followed by satisfaction, civic responsibility, relevance, change agents, a broader appreciation of the discipline and academic integrity. Seventy–eight percent of the participants stated that the experience improved their scientific sampling skills. The impacts on ethics are discussed in the following section.

In this study, subject matter enhancement (i.e. communication skills, broader appreciation of discipline, and societal context) was addressed. Some of the most important assets a college can offer the community are trust, neutrality, and the ability to find common ground among diverse stakeholders. On a practical level, a college can offer a gathering place for diverse groups and individuals from both the private and public sectors to begin a dialogue. The next step is creating dialectic – derived from the Greek word "dialektike." According to Collins, negotiation is necessary to move from environmental impasses to sustainability. [12] The essence of good conflict is using it to reconcile social justice and sustainability in the environmental arena.

Building trust and developing community solutions can be applied in communities on an even larger global scale. The Earth Charter, a document drafted at the Rio de Janeiro conference, emphasized that:

> Human beings are members of an interdependent community of life with a magnificent diversity of life forms and cultures. Embracing the values in the Charter, we can grow into a family of cultures that allows the potential of all persons to unfold into harmony with the Earth Community.[13]

The Charter dictates two principles related to student participation and community involvement. Principle 12 promotes "the participation of youth as accountable agents of change for local, bioregional and global sustainability." Principle 14 ensures "that people throughout their lives have opportunities to acquire the knowledge, values and practical skills needed to build sustainable communities".[14] These principles are consistent with the service learning model, which stresses skills such as negotiation.

Specific texts, institutes, and courses have been established to guide individuals through the fundamentals of negotiation.[15] However, without trust, negotiations and community development are not possible. Negotiation as a strategy to resolve disputes has many advantages over administrative or judicial litigation, mostly because it is cheaper and quicker. Negotiations should be managed by a party or person respected by multiple parties. This is an ideal role for a college. College personnel are often community members themselves, so they are perceived to have a stake in the community. In addition, both community and regulating agencies view faculty and administrators as technical experts. This trust and credibility enhance the dialectic.

SCIENCE EXCLUSIVITY

It is essential that educators reflect on the deficiencies and exclusions in environmental curricula and ponder how to address the shortcomings. Educators need to remember that local knowledge was based on survival so it accounts for uncertainty, resilience, and feedback learning. It also included a respect for nature which is absent from most standardized curricula.

As Gomez-Pompa and Kraus explained after studying environmental educators versed in Western science:

> Many environmental education programs are strongly biased by elitist urban perceptions of the environment and issues of the urban world. This approach is incomplete and...neglects the perceptions and experience of the rural population, the people most closely linked to the land, who have a firsthand understanding of their natural environment as teacher and provider.[16]

In a study on American Indians, Trosper found that there are four commonly (but not universally) held values that are components of respect: community (including views of social obligations and reciprocity); connectedness; concern for future generations (as exemplified by the Iroquois notion of responsibility for the "seventh generation") and humility. In the Gunnison study, participants noted that they gained a broader appreciation of environmental education in its entirety.[17]

ETHICS

Civic responsibility was the fourth highest influenced outcome expressed by the participants after the experience. Related to a sense of responsibility was the participants' high ranking of perceiving themselves as agents of change. Previous service learning research has explored the impacts on students' academic, career, civic, personal, ethical, and social development. Service learning has been shown to positively affect the technical and life skills of participants:

> Service enhances general life skills such as dialogue (i.e., grounded in the understanding that meaning is constructed through an ongoing interaction between oneself and others), interpersonal skills (i.e., the ability to see through the eyes and respond to the feelings and concerns of the other), critical thought (i.e., the capacity to identify parts and the connections among them as coherent patterns and to reflect evaluatively on them, dialectical thought (i.e., the ability to recognize and work effectively with contradictions by resisting closure or by reframing one's response), and holistic thought (i.e., the ability to intuit life as an interconnected whole) in a way that leads to practical wisdom.[18]

This research mirrors the findings in regards to communication, dialectical thought and critical thinking. All participants stated that the experience improved their negotiation communication skills. Ninety-two percent reported improvements in their critical thinking skills.

Environmental education must progress beyond developing environmental literacy to the point of active participation in solving complex environmental issues. An increase in social responsibility is one of the most consistent findings of service learning outcomes.[19] Again, civic responsibility was the fourth ranking in this study. Moreover, service learning research shows that students are more likely to see themselves as connected to their communities, to value service, and to endorse systematic approaches to community problems.[20] This study reinforces previous findings regarding societal context and civic responsibility. As stated previously, the highest ratings in this study showed that service learning influences participants' perception that they had a responsibility to future generations and that one person can make a difference. This is consistent with studies that indicate idealism among younger adults. Dunlap notes, "The tendency of young adults to picture themselves as part of a heroic mission grows out of a normal process of adolescent egocentric thinking, which is the tendency to see oneself uniquely capable of correcting the ills of the world." [21] Conducting a research project that can inform the public and help the community may be more interesting to the students than simple lab work.

QUIETUS

One criticism cited of environmental education is that it often lacks a holistic view of ecological knowledge.[22] This section describes some voices missing from the environmental dialogue, including female and indigenous perspectives. Environmental service learning may hold promise for giving voice to the marginalized, evaluating what we teach and whose voice is acknowledged. Foucault has amply demonstrated that "social power operates through a regime of privileged speakers, having historical embodiments as priest and kings, authors, intellectuals and celebrities. The words of these speakers are taken seriously, as opposed to the discourse of meaningless and often silenced speakers such as women, minorities and children.[23] By excluding certain segments of society, we are teaching students, consciously and unconsciously, what and who is important.

Recent feminist literature promotes linking activism with scholarship and examining the basic power relationships in research. Excluding female voices negates some of the most important attributes identified for deciphering environmental problems, namely, complex and collaborative problem solving.[24]

In October 2004, an accomplishment for women was achieved when Kenyan environmentalist Maathai won the Nobel Peace Prize. She was the first African woman to be awarded the Nobel Peace Prize since it was created in 1901. In the late 1970's, Professor Maathai led a campaign called the Green Belt Movement to plant tens of millions of trees across Africa to slow deforestation. According to Maathai, "it is recognition of the many efforts of African women, who continue to struggle despite of all the problems they face. The environment is very important in the aspects of peace because when we destroy our resources and our resources become scarce, we fight over that.[25] The Nobel Peace Prize Committee says she has combined science with social engagement and politics."

Service learning is a method to credit women for the work in which they participate. Reflection through journal writing is a method to empower women by documenting their stories. It is the cultural space (memory, myth, stories and songs) that constitute the daily life of the community.[26] Native Americans believe story telling instills a gift of power – not a power over but a power from within.[27] Basso's documentation of the close relation between Western Apache stories and the perceivable landscape has already been used successfully in litigation to protect Western Apache land and water rights.[28] Thus, words do equate to power. Implementing service learning projects locally and disseminating the information nationally may inspire future generations of women to create their own sense of community in which their voices are heard.

Recent studies suggest that service stems from the development of empathy for others.[29] If empathy towards others, including non-human creatures, can be developed through service, then it holds promise for educational reform. In conclusion, it is apparent that women possess a strong commitment and valuable knowledge about the environment. However, the challenge is to incorporate their voices into the environmental education system. Gender issues also play into the politics of science including what is deemed acceptable in terms of research, methodology and expert knowledge.

NOTES

1. Orr, 1992
2. Osborn, 1998
3. Ibid, 1998.
4. Daloz, 1997
5. Mendel-Reyes, 1998
6. Manes, 1996
7. Berkes, 1999
8. Ibid, 199.
9. Dewey, 1954.
10. Reardon, 1998
11. Dales, 1997
12. Collins,1998
13. Strong, 1997
14. Strong, 1997
15. Miller, 1989
16. Gomez-Pompa,1992
17. Trosper, 1995
18. Daloz, 1997
19. Giles and Eyler, 1994

20. Ibid,1994
21. Dunlap, 1997
22. Gomez-Pompa, 1992
23. Foucault, 1973
24. Tilbury, 1997
25. Maathai, 2004
26. Shiva, 1993
27. Gray, 1988
28. Basso, 1987
29. Daloz, 1997

Chapter Six: Gown to Town Connection

Only when the last tree has died and the last river poisoned and the last fish has been caught will we realize we cannot eat money.

Cree

A lack of connection to the natural environment not only plagues college campuses, it extends to communities in general. Many segments of modern American society have become culturally disconnected from the natural environment. For example, in Native American traditions, water shaped culture through myth, which, in turn, shaped water management. Modern society's dismissal of cultural values is reflected in our laws, institutions, and natural resources management decisions. Of the 3.5 million miles of rivers and streams, less than two percent are free flowing.[1] Over 5,500 large dams and almost 80,000 small ones impede America's rivers.[2] Therefore, power now shapes water, drowning cultures. Dam construction has displaced almost 80 million people in the last sixty years.[3] If we shift our focus and think of water in legal terms, then the characterization changes dramatically. A good example is the River and Harbors Act of 1899, through which the government claims control of any navigable water. Navigable is broadly defined as any waters that were or could be "made" navigable.

PILOT STUDIES

Present indigenous societies, such as people impacted by the James Bay Hydroelectric Project, offer examples of using local eco-geological knowledge. The community members eloquently describe their relationship to water in different terms than through expert witnesses.[4] The James Bay project used an innovative approach to resource management involving Indigenous Indian tribes and the Canadian government. Indigenous knowledge of twenty-two tribes was used to document changes and traditions based on the local ecosystems. These changes were incorporated as text, and hand-drawn information was compiled using GIS technology.

This knowledge cannot be underestimated. As Berkes states, "traditions are enduring to specific places...Traditions are the products of generations of intelligent reflection tested in the rigorous laboratory of survival. That they have endured is proof of their power.[5] As Worster explains:

> The American West can be best described as a modern hydraulic society, a social order based on the intensive, large-scale manipulation of water and its products in an arid setting that order is not at all what Thoreau had in mind for the region. What he desired was a society of free association of self-defining and self-managing individuals and communities, more or less equal to one another in power and authority." [6]

Rivers hold spiritual value for other cultures as well. Rivers have often been linked with divinities, especially female ones. In ancient Egypt, the floods of the Nile were considered the tears of the goddess, Isis. Ireland's River Boyne, overlooked by the island's impressive prehistoric burial sites, and was worshipped as a goddess by Celtic tribes.[7] However, the temples of ancient India, once dedicated to river goddesses, are now substituted by dams, the temples of modern India, dedicated to capitalist farmers and industrialists and built by engineers trained in patriarchal western paradigms of water management.[8]

For the elders of Eastman at James Bay, the damage done to the river is not an aesthetic or even functional problem. It is a wound to the soul of the river - an offense to its spirit. "For centuries, the concentration of power that comes with controlling water has been a deliberate goal of ambitious individuals, one they pursued even in the face of protest and resistance." How tempting, Worster notes, if you are a politician, to heed only the voices that have the sound of money in them. [9] Yet it is an old American belief that the individual can speak more clearly here than anywhere else in the world. However, water policy is determined by a public agency, which gains power through the exercise of technical expertise, that is, through the reservoirs, dams, siphons and canals it lays out and maintains." Environmental service learning may be a method to heed the voices of community members, younger Americans, and educators. Students can function as conduits in the community. As Mead points out,

> A few committed individuals should not be underestimated when it comes to building communities. [10]

PLACED-BASED WISDOM

Environmental philosophers such as Nabhan promote the idea of incorporating community participation into environmental education. He suggests that one of the most important principles in environmental education, particularly place-based programs, is the teaching from community elders (indigenous or otherwise) about their knowledge of local biota.[11]

As Nabhan explains, environmental education projects should include native interaction in addition to direct experience with the natural world. Environmental education could be enhanced by looking at other cultures and their connections to place and local processes. Oelschlaeger has argued that such relearning is precisely what "wilderness thinkers" such as Thoreau and Snyder attempt to do.[12]

Natives can provide knowledge and role modeling as well. Educators must recognize the participatory mode of experience that indigenous cultures have with nature and learn from that process. Take for example the dreamtime song lines of the Australian Pintupi:

> These song lines correspond to specific routes which can be a source of water, a potential shelter, a high vantage point from which to view the surrounding terrain, or an area of several such clusters. Social taboos, customs, interspecies etiquette --the right way to hunt particular animals or gather particular foods and medicine --all are contained in the Dreamtime songs and stories.[13]

Indigenous cultures use observations in nature to guide their knowledge. As Abram points out:

> The most learned and powerful shaman will be one who has first learned their skill directly from the land itself -- from a specific animal or plant, from a river or a storm-- during prolonged sojourn out beyond the boundaries of the human society. The shaman derives her ability to cure ailments from her more continuous practice of balancing the community's relation to the surrounding land.[14]

A major flaw in our environmental educational system lies in the fact that students are too often confined to texts, labs, and computer screens, which narrows their ability to become observant naturalists. In the words of French phenomenologist Merleau-Ponty, "we must begin by reawakening the basic experience of the world, of which science is the second order expression."[15]

The use of traditional knowledge projects, community-initiated and carried out by aboriginal groups themselves, is perhaps the most common way in which indigenous voices can be heard. Examples of such studies are diverse and include the Darien indigenous land project in Panama;[16] the James Bay Cree trappers' traditional knowledge project in Quebec;[17] the Mushkegowuk Cree land and resource use project in Ontario;[18] and the Marovo Project in Solomon Islands.[19] It is true that educators cannot directly expose each student to indigenous societies or remote environments, but we must acknowledge the shortcomings of our present programs and try to incorporate some natural experience in their educational programs.

In this next section, the importance of a community connection and local wisdom in service learning projects is examined. Service learning programs that place students in the environment can help to combat a loss of contact that many Americans have with nature and other cultures.[20] Plyle describes this loss of contact with nature as the "extinction of experience" and urges educators to address this deficiency through local outdoor programs. Researchers Sward and Marcinkowski found that environmental education efforts that are focused on local areas strengthen the individual's sense of sensitivity and responsibility toward environmental issues.[21] Findings outlined in Chapter Four are consistent with these previous studies. This work is an attempt to combine the necessary skills such as the affective and scientific skills that are needed for educational reform.

Environmental educators are beginning to recognize the power of place in a student's education. Nabhan and Trimble suggest three principles of a place-based program: "Intimate involvement with plants and animals, direct exposure to a variety of wild animals carrying out their routine behaviors in natural habitats, and teaching by community elders."[22] The authors add that educators cut children off from the most important place in their lives and devalue it by emphasizing other places. They argue further that focusing on environmental tragedies can scare children, giving them a sense of powerlessness and cutting them off from the natural world.

Carson an eminent scientist, activist and author, explains just why a sense of natural wonder is so valuable:

> Those who dwell, as scientists or laymen, among the beauties and mysteries of the earth are never alone or weary of life. Whatever the vexations or concerns of their personal lives, their thoughts can find paths that lead to inner contentment and to renewed excitement in living. Those who contemplate the beauty of the earth find reserves of strength that will endure as long as life lasts.[23]

Contemporary educators struggle with ways to allow students to make meaningful connections between academic lessons and their lives. McVey notes that:

> In this age of excessive artifact and hubris, we will go out of our way to engage the wild. To encounter something as strange and compelling as to thoroughly shock us out of our jaded complacency to imagine the world alive again with possibility. To scare hell out of ourselves, if necessary, all to awaken our sense of wonder for a world that perpetually resists our best efforts to understand it. In the wild, we seek the antidote for disaffection. Tonic for a desensitized spirit. Along the way, maybe we discover something about ourselves. Plumb the depths of our own souls, if you will. See whom we are as we look to find where we are. [24]

Harvard biologist, Wilson, has published several articles and books stressing the importance of developing an environmental ethic and sense of place among students. Wilson refers to this as the biophilia hypothesis. Wilson defines biophilia as "the innate tendency to focus on life and life-like processes," noting "to the degree that we come to understand other organisms, we will place greater value on them, and on ourselves".[25] Whether all people are biologically connected to nature is debatable, but it is evident that humans have a link to nature and that relationship varies. Another Harvard professor, Kellert, describes human relationship categories and suggests that these categories are indicative of the human evolutionary dependence on nature as a basis for survival and personal fulfillment. According to Kellert, our classification of values includes the utilitarian, naturalistic, ecological-scientific, aesthetic, symbolic, humanistic, moralistic, dominionistic, and negativistic. These values are so deep that they provide a compelling argument for a powerful conservation ethic. [26] Kellert's classification system illustrates the importance of connections. Service learning projects afford us an opportunity to participate directly in the environment.

Leading environmental writers also discuss the need to cultivate a deep sensory awareness of one's surroundings.[27] Other cultures traditionally have been more connected to the environment. Abrams concludes that the greater connection to nature in native cultures is linked to a more primordial and participatory mode of perception. One component of service learning is the requirement for reflection of experience. As -previously discussed, reflection has been shown in the literature to enhance the learning experience.[28] [29]

Nabhan research indicates that within only a few generations, indigenous cultures such as the O'odham and Yaqui Indians have lost certain ethno biological knowledge.[30] Nabhan raises the question: "How long will regionally extirpated species persist in the oral history, the songs, and the dreams of the O'odham and Yaqui?" Another equally disturbing study indicates that a reduction in direct contact with nature creates a cycle of disaffection and irresponsibility toward natural habitats.[31] Abrams notes: "For it is only at the scale of our direct, sensory interactions with the land around us that we can appropriately notice and respond to the immediate needs of the living world."[32]

FUTURE RESEARCH

The research and professional experience indicate environmental service learning can affect not only students but the environment as well. Further research regarding the relationship between citizenship and environmental science education specifically testing of students' skills rather than measuring their perception of change is recommended. In addition, this study had a small sample size. Future studies could use larger samples and inquire about impacts of the community members not just the students. In addition, the correlation between the four factors such as collaboration, subject matter enhancement, science emphasis and ethics could be developed. The setting was in a rural area so research in urban settings would be interesting for a comparison. A longitudinal study could determine if student's behavior was impacted for a certain length of time. For example, research could determine the extent to which participants change their course of study or professional careers toward "applied" science after service learning experiences. Studies could investigate if by learning about the complexity of environmental problems, students realize they need to be knowledgeable in many different skills, including scientific and communication skills.

State law and international treaties are requiring innovations in environmental education such as community partnerships. Our educational curriculum needs not only to establish such partnerships, but also to train students with a skill base such as negotiation and collaborative thinking, so that these partnerships are successful.

In a competitive global market, graduates need to have a broad appreciation of the discipline. We can no longer produce environmental and biology graduates that do not comprehend the regulatory system that funds their research. The profession requires more than the tradition approach utilized throughout the country. Furthermore, the global economy demands it. It is a disservice to both the students and the environment if educators do not adapt to the changing needs.

PROGRAM IMPLEMENTATION

This final section outlines how to implement service learning in college-level environmental curricula.

For a service- learning program to be successful, the resources of a college must be directed toward the specific needs of a given community. In order to identify the community needs college personnel should host gatherings and attend community meetings. In addition, community leaders should be invited to discuss relations between the college and the community and consider potential partnering projects. Articles should be submitted to local newspapers that explain the principles of service learning and solicit requests for proposals. Proposals could be categorized by subject matter. A master list and project description booklet could be developed and placed in the library and registration office. Another step is to develop a curriculum that is programmatically based, so students can receive credit for their work.

Curriculum options could include:

1. Developing a series of Special Topics courses;
2. Establishing goals for students and partners ;
3. Incorporating environmental service learning components into Biological Sciences, Ecology and Natural History courses;
4. Developing a basic course (i.e., Humanities) to satisfy general education requirements (See Appendix A);
5. Allowing service learning courses to substitute for elective credits.

A key component of all service learning classes is that students ought to reflect on the process and the reasons behind their particular projects. For example, the social and political ramifications of their reclamation projects could be examined. Service learning advocates question whether experience alone will result in help for communities and development of civic consciousness among students . Baker in 1983 called for structured opportunities for critical reflection on services so students "better understand the causes of social injustices and take actions to eliminate the causes."

Another benefit of adding service learning to the curriculum is that students can receive payment for their involvement through the Federal Work Study (FWS) program. All institutions that receive FWS funding are required to spend 7% of their funding on community service. This provides an incentive to add service learning. It also provides a way to address the problems of additional tuition for students. From an administrative standpoint, one way to institute work-study is to link it to community service programs. For example, a Student Employment Handbook Guide might stipulate that eligible employment includes "work in service opportunities ... [including] such fields as childcare, literacy training, environmental planning, environmental education, alternative transportation, and remediation."

The challenge for higher education and financial aid officials is to use the government appropriation on community service. For example, former President Clinton initiated the "America Reads" program and recommended that FWS resources be used to recruit tutors for preschool and elementary students.

Expanding the options for work-study students into other community service areas benefits all partners: the college, students, and the community. The communities gain from the services provided. Work-study students learn about careers in community-based organizations. Placing work-study students in the community is a valuable public relations tool for the college. Finally, on-the-job learning supports most college's workforce development strategic initiative.

In addition to excellent domestic programs, several institutions are now engaged in service learning programs in other countries. An exemplary example is the International Service Learning Experience (ISLE) project that involved sixteen University of Pittsburgh students and an Andean village in Peru. The project consisted of: 1) a seminar on the theory and practice of service learning through the School of Education; 2) bi-weekly, individual reflective writing assignments; 3) communal fund-raising; 4) meetings with professional volunteers and Andean scholars; 5) a series of debriefing exercises and, 6) multimedia documentation and public relations projects encouraging international service learning across campus. The goal was to "produce" global citizens by teaching students about the Andean concepts of reciprocity, or *ayni* .The participants reported a stronger commitment to serve after their experience. [33]

Incorporation of service learning into the biological sciences has been gaining ground since it was promoted through the Corporation for National Service's SEAMS program. The first SEAMs report, "Science and Society: Redefining the Relationship", clearly displayed the interest many science professors have in bringing their scientific expertise closer to public schools, health agencies, and other community organizations.

Further research needs to be conducted about environmental service learning to determine if both the community and students can benefit by working collaboratively on local environmental issues. According to Orr, education relevant to the challenge of building a sustainable society will enhance the learner's competence with natural systems. Orr proposes that:

> Lead institutions should conduct research to support sustainable livelihoods, sustainable communities and ecological economies and link research and service to wider community efforts to establish just and sustainable cities, bioregions, and global economies.[34]

According to Clugston, a university leader in the sustainability movement, "while there are hopeful signs that academic disciplines and institutions are responding to the environmental challenges, outreach is not present in most academic settings."[35]

As part of this redefinition, national legislation needs to reflect teaching methodologies such as service learning. California passed bill AB 1548 – that establishes a pilot program to promote service learning partnerships. California has historically implemented environmental regulations that are more stringent than federal standards. Other states tend to monitor California's legislative initiatives and implement similar state programs. If the pattern follows for educational regulations, elementary and secondary school districts across the nation will be required to implement a model environmental curriculum.

AB 1548 created a new agency, The Office of Education and The Environment (OEE). The OEE was required to develop a model curriculum and to make the curriculum available electronically. This curriculum imposes environmental educational principles. The basic elements include but are not limited to (California Educational Code, Section 33541):

(1) Integrated Waste Management
(2) Energy Conservation
(3) Water Conservation and Pollution Prevention
(4) Air Resources
(5) Integrated Pest Management
(6) Toxic Material
(7) Wildlife Conservation and Forestry

CONCLUSION

Environmental service is an important key to addressing environmental issues within communities because public and private agencies tasked with environmental duties are generally under-staffed and under-funded.

Educational institutions should be more proactive in the process. By participating in community projects, students may become more socially responsible and gain recognition of the interdependence of members (human and non-human), organizations, and institutions within a community.[36] In addition, Kelly points out that, "isolation from local community members and problems teaches students to ignore those who are not directly related to their professional advancement".[37]

Another improvement advocated by environmental professors is utilizing place-based ecology and a bioregionalism approach. Place-based ecology includes the study of human and ecological communities, history and natural sciences. The bioregionalism approach emphasizes place and community as a means to construct broader approaches to environmental policy.[38] Achieving a reflective sense of place may also contribute to an ethic of caring about habitats and communities.[39] One method to address place-based ecology and the identified professional skills mentioned above is through the implementation of an environmental service-learning course. If services learning can promote a deeper connection to a particular place, which research has shown to be true, then one can hypothesize that a person's commitment to their local environmental challenges will be positively impacted.

Environmental service learning can be an approach to achieving the goal of "a state of harmony between humans and land".[40] By combining scientific knowledge and ethical judgments, our educational institutions could lead us toward a more sustainable society.

> Whatever attitude to human existence we fashion for ourselves, know that it is valid only if it be the shadow of an attitude to nature...The ancient values of dignity, beauty, and poetry which sustain us are of nature's inspiration...Do not dishonor the earth lest you dishonor the spirit of man.[41]

To summarize, this study identifies key aspects of the student's environmental service learning experience and theorizes how these outcomes relate to critical environmental science skills. Participants ranked in order of importance the benefits of adding a societal context to their scientific work, satisfaction, learning to cooperate, developing a civic responsibility and a broader appreciation of the discipline as the most important outcomes of the experience. Service learning is a pedagogy that combines the affective and scientific skills necessary for producing environmental graduates with the skills needed to meet the current environmental challenges facing society. Participants enjoyed the experience, which may lead to continued involvement outside the educational institutions. Service learning is a method to connect the colleges with the communities and not isolate them. This information can assist in enhancing environmental education. As Parker, a leader in education transformation, stated, "When you realize that you can no longer collaborate in something that violates your own integrity, your understanding of punishment is suddenly transformed." [42]

This environmental empowerment study enlarged the service learning research specifically by conceptualizing its potential within the environmental education field. It presents pedagogy relatively new to the science field. This approach may hold potential for linking the affective and cognitive skills. The affective skills such as behavior, ethics and emotional quota are skills recently found to be the critical for successful leaders.[43] Certainly, the planet needs strong environmental leaders. Our education system needs a transformation to inspire leaders and assist communities. This research centered on balancing the environmental justice scales by examining how power, education, and the legal system affect the environment. The study also documents methods to increase the number of minorities and women in the science fields. Finally, the research suggests important environmental education curriculum changes pursuant to recently passed legislation and critical needs.

NOTES

---

1. Michael, 2000
2. Ibid,2000.
3. Worster, 1995
4. Ettenger, 1993
5. Berkes, 1999
6. Worster, 1995
7. McCally, 1996
8. Shiva, 1985
9. Worster, 1995
10. Mead, 1985
11. Nabhan, 1994
12. Oelschlaeger, 1992
13. Payne, 1989
14. Abram, 1996
15. Merleau-Ponty,1962
16. Gonzalez et al, 1995
17. Bearskin et al., 1989
18. Berkes et al. 1994, 1995
19. Baines and Hviding 1993
20. Plyle, 1993
21. Sward, 1997
22. Nabhan and Trimble, 1994
23. Carson,1990
24. Mo, 1998
25. Wilson, 1993
26. Kellert, 1993
27. Abrams, 1996
28. Eyler and Giles, 1999
29. Gray, 1999
30. Nabhan, 1993
31. Pyle, 1992
32. Abrams, 1997
33. Porter, 2001

34. Orr, 1998
35. Clugston, 1998
36. Kelly, 1998
37. Ibid, 1998
38. McGinnis, 1999
39. Thomashow, 2002
40. Leopold, 1966
41. Beston, 1993
42. Parker,1998
43. Goleman, 2002

APPENDIX A

Instructor: Dr Rosemarie Russo
Classes: August 28, Sept 11, Oct 16, Nov 20, (2:00 – 3:00);
Mount Antero Conference room – Administration Building
Office Hours: Friday 12:00 p.m. - 1:00 p.m. or upon request.
Office Phone Number: 204-8113
Email: drperseverance@yahoo.com

Course Title: Biology 194: Service Learning: Environmental Engagement

"Only when the last tree has died and the last river poisoned and the last fish been caught will we realize we cannot eat money." -Cree Elder

"This we know. The earth does not belong to man; man belongs to the earth. This we know. All things are connected like the blood, which unites family. All things are connected. Whatever befalls the sons of the earth? Man did not weave the web of life; he is merely a stand in it. Whatever he does to the web, he does to himself." -Chief Seattle

"UNLESS someone like you cares a whole awful lot, nothing is going to get better. It's not."
-The Lorax, Dr. Seuss

Course Description

Bio 194 aims to educate on the philosophical, educational and scientific aspects of environmental protection. Its purpose is to empower students and citizens with practical, methods for advocacy and sustainability.

Human communities and the environment are increasingly endangered in today's society. Pressures for economic growth, the world economy, and the expansion of state and national regulations have decreased local control and have heightened the strains on the environment. This class will examine these conditions at a local level and see how a community can strengthen its environmental stewardship. An exploration of the connection between basic human well being and a healthy environment will be explored.

What are earth rights? Is there a need for a new environmental ethic? How can community members and students coordinate campaigns to improve political efficacy and save the planet? What institutions protect and promote human/earth rights? These are just some of the core questions we will explore together.

Through readings, class discussions, guest speakers, videos, role playing, grassroots and activism, educational excursions, soul searching and journal writing, one will gain a better understanding of environmental sustainability. One can also begin to establish a person philosophy and human rights framework to analyze issues and engage in personal activism.

II. Course Objective

The course short-term goals are to promote the practice of environmental education for all ages and abilities, environmental stewardship, and economic revitalization based on environmentally sound business practices

The long-range goals and objectives are to expand its efforts along important fronts: watershed protection, educational outreach, sustainable industry, sustainable visioning, sustainable agriculture and landscape ecology, sustainable energy, sustainable forestry, and environmental planning.

Projects will be undertaken to help develop understanding of ecological ethic to form opinions on issues of law, science, politics and morals, assess the impact of education on developing a health community and campus, explore ability of self empowerment from ethical position to activism through education and experience; trace the historical development of law connecting the environment to human rights; and understand the interconnectedness of earth rights and human rights; respect indigenous wisdom in relation to living in harmony with the earth

Each project will have different skill competencies but for the Interpretative trail, an objective would include: To further an appreciation for the natural history of our world, while concentrating on plant communities in selected Colorado habitats. The purpose is not to learn all the plants available to us, but rather, to learn how to identify them and to recognize some basic plant communities. A second skill is to teach and guide others in the concepts of wildflower identification and conservation.

Course Requirements

Grading

Your grade will be determined as follows:

A = 93-100	**C+=78-79**
A- = 90-92	**C= 73-77**
B+ = 88-89	**C-=70-72**
B = 83-87	**D=60-69**
B- = 80-82	**F=Below 60**

Journal – 20%

Postings – 30%

Community work – 30%

Final Project/Paper/Website – 20%

Written Assignments: Letter grades for assignments equate to the numerical. You are required to submit a total of 15 written assignments (4) journal assignments, (9) postings and (2) responses to classmates and either a paper/website/or project description

Project(s)

Students can choose from the three different projects or work on a select project. In order to receive an Americorp stipend – 300 hours need to be completed by the end of the spring term. In order to receive credit a total of 54 clock hours need to be completed. The projects include:

Invasive Species Project (Larimer County): Invasive Species Removal Project involves the removal of salt cedar and Russian Olive trees in order to promote biodiversity and conserve the natural ecosystem. The baseline database will be shared with Land Management agencies in the county. It will also include finding an artistic use for the cut material. Another component is the production of ancillary materials for presentations, promotions, funding requests and educational training. Three (3) students can be funded through AmeriCorps.

Community Garden Project (Johnson/Front Range Campus) This project will plan, design and construct a community garden on campus with special attention to native species and vegetables. Produce will be given to community groups.

Interpretive Trail (Front Range Campus) The service-learning component replaces the "typical term paper." The on-campus component will concentrate on the campus location. Others will work with 3rd graders at the Johnson Elementary School, to teach them skills in wildflower and native plant identification, and then to help them implement a nature trail, which will be accessible to the community. Your own learning should be enhanced through the teaching of your new skills to others.

You are expected to participate fully in class workshops, as well as to make commitments on your own time to meet with small groups of Johnson Elementary students to help them with their designated portion of the nature trail. Finally, you are expected to evaluate your own participation in the service-learning project.

Students can also choose from a list of projects outlined in the service learning manual or choose a project of their own with the professor's approval.

Journal (20 points)

An important part of service-learning is reflecting on one's activities. Each of the students will keep journals. The instructor will review these each month for four times. The content of the journal will vary from student to student and from week to week. The primary intent of a journal is to provide the students with the opportunity to reflect on what you have learned. This may include making observations, asking questions, expressing feelings, or documenting information. The journal can also be used to record information such as interviews, data gathered in the field or directions as to the next step to take on a project. There is no set rule for the appropriate length of a journal. If there are doubts, about whether you are doing enough, give an instructor an entry or two and ask for feedback.

A journal/notebook will be kept during all class field trips and your own explorations. Include class field notes, work with students, sketches, descriptions of plants, habitats, micro-habitats, family characteristics, plant associations within communities, variations within species, seasonal progression, attitudinal changes, dates and locations of specimen collections, poems, etc. It should be as detailed as possible and although your journal will not be graded for artistry, you should feel free to be creative.

Reflection /Postings (30 points)

After each experience with your group, please write down your impressions. How is the project progressing from your perspective? What suggestions might you make for yourself or for others for future meetings? Does your work with the student's influence how and to what extent you learn the material? If so, how? Would you be learning the material quickly if you were not involved with this project? Please be as detailed possible. What worked well? What didn't work so well?

Trail /Garden Work Log Document (30 points)

Example of Trail Work:
Take field guide, notebook, markers, flagging tape and stakes with you;

Remind students to bring notebooks, guides, plant press and magnifiers;
Keep a running list of all plants identified;
Try to be selective on what you identify -- (there is no need to ID every trout lily);
Get the students back in time for their next activity;
Turn in your plant list to the "list organizer" who will coordinate as master list of plants; and pin-point as best you can on the map any less common plants.

Final Project/Website or Paper (20 points)

All students are expected to do a final presentation/or paper/website related to the project. This may be done individually or in small groups, or by all the interns. These may take a wide range of formats. For example, it is possible to do a photo essay of a particular problem or issue. You may choose to write a section of a grant to the National Science Foundation. You may prefer to conduct a series of interviews on a topic and to transcribe these interviews into a document. The options are almost limitless. The key factor is that the work provides you with an educational opportunity and that the results contribute in some way to the goals of the project. These are graded as a paper or project would normally be graded. Let me know if your schedule changes.

Modules

There are four modules to this course. The first is to develop greater communication and interaction between the partners noted. The second is to provide you with a hands-on opportunity to help the community address a problem and work inter-disciplinarily in formulating and addressing the problem. In this course, you go beyond identifying environmental problems to actually focusing on solutions. This takes two forms: (1) Who: becoming familiar with, learning from, and celebrating individuals and groups who have themselves achieved success at solving environmental problems, and who are thus role models or examples for others to follow or emulate, and (2) How: as part of a 2-5 person interdisciplinary consulting group/team/task force, applying problem-solving skills to real-world environmental problems and thereby helping the community. Third, it will integrate service learning not only across disciplines but also across academic boundaries and into the community. The final module is to establish the foundation for addressing the issues of the environment in a more in-depth and ongoing manner.

Students will play a critical role in determining the activities to be undertaken, the process in which issues are addressed, and doing the actual work. You are expected to attend and actively participate in all the described activities and are responsible for all announcements made during those activities. Active participation is expected, and participation points will be lost for poor attendance with unexcused absences, poor performance regarding journal entries or letter to a member of Congress.

College students examining global environmental issues are often overwhelmed by the enormity of the problems and feel that the issues are so complex that there is little or nothing they can do about them. This course attempts to empower you to make a difference, to do something about the global environmental problems, by acting at the local level...but only after you have first focused on where you want to be as an individual and member of society and how you want to get there. To that end, we examine change strategies for creating sustainable communities, for bridging the gap between our utopian visions and the present-day realities, for getting from where we are to where we would like to be in order to overcome the feelings of lack of control, despair, paralysis, apathy, inaction, doom and gloom, confusion, powerlessness, and helplessness by having you think globally and act locally. Action at the local level is often the first step toward a solution to global environmental issues.

Class Format:

The course is conducted as a seminar. Group's discussion of the readings and assignments constitutes the structure for most of the class discussions, although there are occasionally lectures, guest speaker, individual and group presentations and off-campus trips. Active participation by all class members (including raising questions related to reading assignments) is essential for the class to be a success.

Readings

To derive maximum benefit from the course, it is essential that you: (1) complete the assigned readings in advance of class; and (2) allow time for reflecting on what you read (journal entries will help in this regard). Some readings on reserve in the library or available as photocopied handouts offer advice and numerous problem-solving techniques to assist your group:

Projects:

The project(s) provides a real-world opportunity to apply problem-solving skills to a complex (higher-order) environmental problem by a deadline --thereby gaining a better understanding of that issue-- and to help a real client (agency, institution, organization or business) in need. This hands-on, learn-by-doing project will hopefully strengthen all four skills involved in effective problem solving (creative thinking, critical thinking, intelligible management, and good communication) and will help you integrate them with better results.

Problem-solving techniques are presented in class and readings --by the instructor, guest speakers, and student teams; in addition, in class you are presented with hypothetical situations in which you apply what you are learning about problem- solving and other important job-related skills (e.g., working in groups, facilitating effective meetings, problem identification and definition, management skills, organizational skills, time management, writing a group document, making oral presentations) are offered every semester; others, as they are needed by groups to complete their projects.

Topics are chosen to provide relevant field experience and --most importantly-- to be of real value to the community - As you apply techniques to environmental problems as professional environmental consultants (interdisciplinary consulting groups/teams/task forces), I will emphasize process and approach --how you are doing what you are doing and why; and will offer a generalized methodology for solving environmental problems for your team to utilize. I will devote much class time to helping you with your project, and expect a high quality, genuinely useful, professional product --delivered on time. Members of each group share the same grade for their work. During the last two weeks of the class, each group makes an oral presentation, submit a project (poster board), or post a website with an explanation to the rest of the class and to clients, and each member must participate in some way in that presentation. Clients and community members are invited to the presentations

Several conference assignments (most employing application of problem-solving techniques to your problem) help keep you on schedule. Following is the period, within a 13-week semester:

Week 2 (Conference 2): Identification of project topic and team membersWeek 4 (Conference 2): Preliminary Outline
Week 5 (Conference 4): Congress Letter
Week 7 (Conference 4): Presentation of "mock-up" of the desired project and gaining feedback from others

Week 11 (Conference 5): Completion of first draft of report/website/project
Week 12: (Conference 6): Completion of final, polished website, draft of report; rehearsal of oral presentation.
Weeks 13 (Conference 7): Oral project presentations in class; critiques by fellow course participants and instructor(s); evaluation of process; evaluation by client.

Texts

Required:

Stanton, T. (1999) *Service Learning*. San Francisco, CA: Jossey-Bass Press.

References:

Guennel, G. (1995) *Guide to Colorado Wildflowers*. Englewood CO.: Westcliffe Publishers, Inc.

Earthworks (1991) *The Student Environmental Action Guide*. Berkeley, CA.: Harper Collins Publisher.

Kriesberg, D. (1999) *A Sense of Place*. Englewood Colorado: Teacher Idea Press

Lantieri, L. (2001) *Schools with Spirit*. Boston, MA: Beacon Press.

Palmer, P. (1998) *Courage to Teach: Exploring the Inner Landscape of a Teacher's Life*. San Francisco, CA: Jossey-Bass.

Light, R."Diversity on Campus," Chapter 7, pages 129 159. In Light, Cambridge, Massachusetts: Harvard University Press, 2001.

Appendix B
Service-Learning Agreement Form
Student/Agency

Student Name: ______________________________

Phone Number: __________________________

Class Name: ________________________________

Instructor's Name: ________________________

Name of Organization:

____

Supervisor or Coordinator: _____________________________

Phone Number: __________________

Service-Learner Job Description:

____

____

____

Student: I agree to...

☐ Perform my respected duties to the best of my ability.

☐ Adhere to organizational rules and procedures, including record-keeping requirements and confidentiality of organization and client information.

☐ Be open to supervision and feedback, which will facilitate learning and personal growth.

☐ Complete ____ hours of service per week from the time beginning _____ (mo), _____ (day) and ending _____ (mo), ______ (day). If specific days and hours are agreed upon, they are listed as follows:

M_____ T_____ W_____ Th_____ F_____

☐ Meet time and duty commitments or if I cannot attend, to provide 24 hour notice so that alternative arrangements can be made.

Supervisor: I agree to...

☐ Provide adequate information and training for the service-learner including information about the organization's mission, clientele and operational procedures.

☐ Provide adequate supervision to the service-learner and provide feedback on performance.

☐ Provide meaningful tasks related to skills, interests and available time.

☐ Provide appreciation and recognition of the service-learner's contribution.

Student ____________________________________ Date ___________________________

Supervisor ______________________________ Date _____________________________

Adapted from The Service Integration Project – C.S.U.

AUTHOR BIOGRAPHY

Dr Rosemarie Russo currently teaches Global Environmental Change for Thomas Edison State College, law at the Graduate School of Education, Colorado State University and Sustainable Practices at University of Colorado – Boulder. She serves as the Sustainability Coordinator for the City of Fort Collins. She has been a professor, Environmental Chair and Dean for Athletics, Biological and Oceanography Division at DVC. She has worked as an environmental law consultant on Superfund sites. Rosemarie has taught abroad in Central America. She earned a BS in Environmental Science from Rutgers University, an environmental Masters degree from Vermont Law School and a doctoral degree from Argosy. She is a certified comprehensive science teacher and a volunteer backcountry ski ranger at Cameron Pass.